HUMAN-ISH

What TALKING to YOUR CAT or NAMING YOUR CAR REVEALS about the UNIQUELY HUMAN NEED to HUMANIZE

JUSTIN GREGG

Little, Brown and Company

New York Boston London

Copyright © 2025 by Justin D. Gregg

Hachette Book Group supports the right to free expression and the value of copyright. The purpose of copyright is to encourage writers and artists to produce the creative works that enrich our culture.

The scanning, uploading, and distribution of this book without permission is a theft of the author's intellectual property. If you would like permission to use material from the book (other than for review purposes), please contact permissions@hbgusa.com. Thank you for your support of the author's rights.

Little, Brown and Company
Hachette Book Group
1290 Avenue of the Americas, New York, NY 10104
littlebrown.com

First Edition: September 2025

Little, Brown and Company is a division of Hachette Book Group, Inc.
The Little, Brown name and logo are trademarks of Hachette Book Group, Inc.

The publisher is not responsible for websites (or their content) that are not owned by the publisher.

The Hachette Speakers Bureau provides a wide range of authors for speaking events. To find out more, go to hachettespeakersbureau.com or email hachettespeakers@hbgusa.com.

Little, Brown and Company books may be purchased in bulk for business, educational, or promotional use. For information, please contact your local bookseller or the Hachette Book Group Special Markets Department at special.markets@hbgusa.com.

Print book interior design by Bart Dawson

ISBN 9780316577588

LCCN is available at the Library of Congress.

Printing 1, 2025

LSC-C

Printed in the United States of America

*To my daughter, Mila, who loves rats and snakes
and makes me proud*

HUMANISH

CONTENTS

Introduction. 1
Let's Talk Fauxnads: A Crash Course in Anthropomorphism

Chapter 1. Fur Babies 17
Why We Treat Pets like People

Chapter 2. Dolphin Doulas 43
How Anthropomorphism Makes Scientists Crazy

Chapter 3. Wally the Alligator 67
Why We Misunderstand Ugly Animals

Chapter 4. Stove Spiders. 89
How and Why We Don't All Anthropomorphize the Same

Chapter 5. Companion Cube. 115
Why We Get the Feels for Inanimate Objects

Chapter 6. Creepy Counterfeits 143
When Humanish Things Get a Bit Too Human

Chapter 7. AI Overlords 169
How We Anthropomorphize the Incorporeal

Chapter 8. Cute Capitalism 201
How Marketers Use Anthropomorphism to Manipulate People

CONTENTS

Chapter 9. Puppy Propaganda 225
How Anthropomorphism Helps Us Kill People

Conclusion 243
*How to Harness Anthropomorphism to Create
Healthier Relationships*

Acknowledgments 253
Notes 257
Index. 285

INTRODUCTION

Let's Talk Fauxnads
A Crash Course in Anthropomorphism

When Gregg Miller's dog disappeared for three days, Miller reluctantly agreed to remove his testicles. The dog's testicles that is, not Miller's.

Buck, the dog, had wandered off looking for a female to mate with, perfectly normal behavior for a male dog with intact testicles. To stop sexually mature dogs like Buck from roaming too far, most veterinarians recommend having the dog neutered. But Miller, like many men, struggled with the idea of canine castration. "It's become our culture to accept emasculation of our pets as normal," Miller told CTV News. "Turning our pets into little eunuchs."[1]

Despite his misgivings, Miller allowed Buck to be neutered. But after the surgery, Miller was heartbroken watching Buck lick mournfully at his recently emptied scrotum. "[Buck] was telling me, 'They're gone. What happened?'" recounted Miller.[2]

It was in that emotionally charged moment that a solution—and a killer business idea—struck Miller: prosthetic dog testicles.

In response to Buck's perceived loss of dignity, Miller invested his savings in developing fake dog testicles. His initial prototypes—which he dubbed Neuticles—were made of hard plastic. They could be inserted into the scrotum of a neutered dog to re-create the look and feel of testicles. The surgery itself takes only a few minutes and is "as easy as changing a lightbulb," according to Miller.[3] Early adopters complained that the testicles made a disconcerting "clonking" sound when their dogs jumped onto the couch. Eventually a quieter, silicone version was created, and these took the world by storm. Miller would go on to dominate the market in fake dog balls. To date, Miller's company CTI (Canine Testicular Implantation) has sold over 500,000 Neuticles (also called "fauxnads"), earning him millions and an Ig Nobel Prize (a satiric award given to scientific research that makes people laugh, then think).

Miller's crusade was predicated on the desire to help neutered dogs retain their "natural look and self-esteem," as the Neuticles marketing material suggests.[4] But is it *true* that dogs get sad when they notice their testicles are missing? Do they even notice in the first place?

"Yes, they do!" insists Miller.[5]

Not so, say many animal cognition scientists—myself included. The vast majority of veterinarians see no physical or psychological benefit for the animals' getting testicle implants, and most are unwilling to perform the surgery. The Royal College of Veterinary Surgeons, in the United Kingdom, states that "the insertion of prosthetic testicles is not a procedure that benefits the animal and is not in the animal's interests."[6] The Veterinary Council of New Zealand warns that "the insertion of neuticles

INTRODUCTION

(prosthetic testicles) cannot be justified."[7] Veterinarian Joe Dunne pulled no punches in lambasting the idea of "post-neutering trauma," writing that "there is no medical or psychological reason why a dog would need false testicles. All perceived benefits are for the owner's benefit, which makes their implantation an unnecessary and selfish procedure."[8]

So what is going on here? Why, despite objections from experts, does Miller feel so strongly that his dog Buck experienced a personal crisis after castration? Why did my good friend Wojtek shed tears—actual *tears*—after getting his dog neutered? "I wept inconsolably in my car after dropping my dog at the vet," he told me. "I don't remember when I mourned so intensely."

All this drama around dog emasculation rests at the furry feet of anthropomorphism.

Anthropo-what-now?

This is a book about anthropomorphism and the powerful grip it has on your mind. So powerful that it makes grown men weep in their cars at the thought of canine castration. It's also about the pleasures of anthropomorphism, and how it can be harnessed as a force for good in your life. Anthropomorphizing your cat or your car or your AI companion, if, when done thoughtfully, is both fun and beneficial to all parties involved.

If you aren't familiar with anthropomorphism, allow me to introduce you to the topic. Prepare to be amazed as I unveil a world filled with emotional support alligators, a woman who married her briefcase, and Soviet superbabies who drink dolphin milk. If you *are* familiar with the word, that's fantastic too because you're probably wondering how dolphin milk fits in and I can't wait to tell you.

But before I explain why anthropomorphism makes people so weird about their dogs' testicles, I need to make sure we're all on the same page since the word *anthropomorphism* can mean different things to different people. When I told my friend Nathan and his wife, Clare—she is a rector in the Anglican Church—that I was writing a book about anthropomorphism, they were slightly puzzled. It was weird that I, an openly nonreligious animal cognition scientist, was interested in the topic. For a religious scholar, the word *anthropomorphism* applies to the problem of depicting god(s) in human form, and they were less familiar with it as a concept that applies to the study of animal behavior. When I was chatting with the evolutionary biologist Marc Bekoff about my book idea, he was surprised when I told him that I was going to dedicate a section of the book to artificial intelligence (AI). He'd spent an entire career using the word *anthropomorphism* to describe how humans interact with and study animals, and hadn't considered how it might apply to AI. And when I asked a few Generation Alpha kids if they knew what the word meant, they told me that anthropomorphism is what made their gamer friend simp over a chatbot. They'd never heard it being associated with either animals or gods.

Throughout my training as an animal cognition scientist, *anthropomorphism* was a well-known—and rather infamous—dirty word, a potential pitfall that could fool us into thinking that animals had human-like cognitive traits where there were none to be found. The battle between scientists over the use and/or misuse of anthropomorphism as a tool for understanding animal minds has been raging since Darwin's time, as you'll read about in future chapters. But you won't find the word *anthropomorphism* used to refer to animals in very many news stories these days. Most headlines use the word to describe the problematic way that people

INTRODUCTION

unjustly attribute human qualities not to animals but to AI. "The Perils of Anthropomorphizing Machine Intelligence," reads one headline from a 2023 *Newsweek* article.[9]

A modern definition needs to capture the diverse ways in which we use the term today, and it needs to fit just as well for animals and gods as it does for AI. The behavioral scientists and anthropomorphism experts Adam Waytz, Nicholas Epley, and John Cacioppo offer this definition of anthropomorphism: "attributing capacities that people tend to think of as distinctly human to nonhuman agents, in particular humanlike mental capacities (e.g., intentionality, emotion, cognition)."[10] I like this definition, but for the purposes of this book I don't want to limit things to "agents," as an agent is some sort of being that has the ability to act or do things, and many things we anthropomorphize, like pet rocks, don't *do* anything at all. The Cambridge Dictionary's current definition casts a wider net, defining *anthropomorphism* as "the showing or treating of animals, gods, and objects as if they are human in appearance, character, or behaviour." This definition includes the flip side of anthropomorphism, which is that it not only nudges us to see humanity in the world around us but also leads us to *depict* the natural world in human-like form in our art and literature. Think the Ents (human-like tree creatures) from *Lord of the Rings* or Winnie-the-Pooh (a human-shirt-wearing but otherwise animalistically pantsless talking bear). In addition to agents (like animals, gods, or pantsless bears) and non-agential objects (like cars), I would include natural events (like hurricanes), abstract concepts (like corporations), and computer software (like AI assistants) on the list of things we anthropomorphize. Anthropomorphism is then perhaps best defined as this: the human propensity to treat or depict everything that exists as if those things were human.

Now that we have a handy definition of anthropomorphism on the books, it's time for a crash course in how anthropomorphism works. This quick overview will provide you with the tool kit you'll need to make sense of the madcap examples of anthropomorphism that are coming down the pike. To get started, we have to agree on what a "human" is and decide what it means to "treat" something "like a human." Figuring this out is not just an exercise in pedantry or *mierenneuken* (the delightfully vulgar Dutch term for nitpicking that I only feel comfortable translating in the endnote).[11] Rather, it's a necessary step on the way to explaining how and why our minds generate anthropomorphism.

Let's start with the question of what a human is. "Human" is a biologically ingrained category our minds generate for other members of the species *Homo sapiens*. Our biology and our genes are primarily concerned with keeping us alive long enough to have children, so it's not surprising that our minds are experts at differentiating between humans and not-humans. That way, we don't waste our time mating with or spending too much time worrying about the welfare of non-human entities that won't help us pass on our genes. This kind of intra-species recognition is a fundamental aspect of all animal species' biology. As much as a chimpanzee, for example, looks and acts like a human (they share almost 99 percent of our DNA, after all), a normally functioning human brain will automatically classify a chimpanzee as not-human. Humanish, to be sure, but not an actual human. This ability to effortlessly differentiate between human and not-human is key to understanding how anthropomorphism works (especially when it comes to anthropomorphism gone wrong), as you will see throughout the book.

As far as the question of what it means to "treat" something "like a human," this is where things get messy. To treat something

INTRODUCTION

like a human means to interact with it in a manner that assumes it has a human-like mind. Or more accurately, to treat it as if it has a mind similar to your own. As a human, you have a mind filled with emotions, desires, thoughts, beliefs, etc. So to anthropomorphize something means to interact with it in a way that assumes that it has its own set of emotions, desires, thoughts, and beliefs that are analogous to yours.

Human minds are stuffed with cognitive abilities that can be lumped into two broad categories: experience and agency. In Daniel Wegner and Kurt Gray's book *The Mind Club*, experience is described as involving the capacities for "hunger, pain, pleasure, rage, and desire, as well as personality, consciousness, pride, embarrassment, and joy," whereas agency involves "self-control, morality, memory, emotion recognition, planning, communication, and thought."[12] To treat a fellow human as a human means to acknowledge that they have a mind chock-full of cognitive abilities divided across these two domains, and this means we should grant them some sort of moral status in deference to those abilities. It doesn't mean we necessarily always feel the urge to be nice to every human we encounter, however. If some jerk in a Porsche is riding our bumper on the highway, we might slow down and take pleasure in watching them through the rearview mirror as they lose their mind with road rage. Sometimes we cause harm to others precisely because we know they have a mind that can generate experience and agency similar to our own. But, as we will see throughout the book, once we lump someone in the category "human," our minds generate a qualitatively different moral experience when thinking about them as opposed to anything in the non-human category. As Adam Waytz points out in his book *The Power of Human*, humans "respond with moral concern to the slightest indication of humanity."[13] Importantly, this distinction between experience and agency and the way it interacts with our capacity for moral concern will help explain why we anthropomorphize different things differently.

So anthropomorphism, then, is the act of treating a nonhuman as if it had a human-like mind even though we *know* it is not a human. And this weird, irrational stance is precisely why people have been shitting on anthropomorphism for the past two and a half thousand years. The sixth-century BCE Greek philosopher and poet Xenophanes mocked those who pictured the pantheon of gods as a bunch of humanoids lounging around Mount Olympus dressed in the latest Grecian fashion, writing that "mortals suppose that the gods are born (as they themselves are), and that they wear man's clothing and have human voice and body."[14] The seventeenth-century philosopher Francis Bacon, often viewed as the father of empiricism (the idea that knowledge comes from observations, measurements, and experiments), was worried that humans tend to view not just the divine but the natural world as a whole in inappropriately human terms. Bacon wrote that "human understanding is like a false mirror which receives light irregularly, then distorts and discolors the nature of things by mingling its own nature with it."[15] Eighteenth-century philosopher and empiricism fanboy David Hume felt the same, writing that there is a "universal tendency among mankind to conceive all beings like themselves, and to transfer to every object, those qualities, with which they are familiarly acquainted, and of which they are intimately conscious."[16] George Henry Lewes, a contemporary and critic of Charles Darwin, cautioned against assuming there was a human-like mind motivating the behavior of non-human animals, writing in 1858 that "we are incessantly at fault in our tendency to anthropomorphise, a tendency which causes us to interpret the actions of animals according to the analogies of human nature."[17]

INTRODUCTION

So if anthropomorphism is such a problem, as these prominent thinkers believe, why did it evolve in the first place? Why is it a universally observed behavior, part and parcel of the human condition across cultures and throughout history? What's causing us to treat clearly non-human things as if they are humans? Shouldn't our biology and genes prevent us from anthropomorphizing, in much the same way that they stop us from trying to make babies with capybaras? As I hope to make clear in this book, there's a good argument to be made that being predisposed to find human-like qualities in the world is not a silly bug but a vital feature of the human mind.

But why, though?

There are two main hypotheses as to the biological benefits of anthropomorphizing.[18] The first is that anthropomorphizing non-humans prevents us from missing an opportunity to connect socially with another human. This is called the social motivation hypothesis (the drive to seek out human minds in the world). If our default setting is to assume that there are human-like minds in all things—and especially things that appear vaguely human—then it guarantees that we'll be correct when we encounter an actual human mind. And connecting socially with another human through a mutual understanding of shared subjective experiences is the single most powerful thing that our species can do. It allows us to cooperate and to coordinate our behavior—in terms of both complexity and scale—in ways that no other species can. Every human relationship, whether that's with our spouse, our pickleball partners, or the millions of other people with whom we share a nationality, is predicated on our unshakable belief that we all possess similar minds. We might not see eye-to-eye with our neighbor when it comes to specifics like politics or choice in music, but these are just surface differences. As Martin Luther King Jr.

wrote, "We are caught in an inescapable network of mutuality."[19] And that mutuality derives from the identical cognitive architecture of experience and agency upon which all human minds are built. Anthropomorphism, then, evolved to guarantee that we'd never miss the opportunity to network with a like-minded other human. There's a lot of evidence to support this hypothesis, as we'll see in this book, including the simple fact that humans deprived of social contact (i.e., lonely people) tend to anthropomorphize more than socially active people.[20]

The second hypothesis as to why we anthropomorphize non-humans is that it helps us make sense of an unpredictable world. This is called the effectance motivation hypothesis (the drive to explain observable behavior via human-like intentions). When we are young children, the world appears full of unfamiliar and confusing things and events. The one thing that we have the most exposure to—and thus the most familiarity with—is our own thoughts and feelings. So as we grow and try to make sense of the newness in the world around us, we use our own minds as the model to explain the unpredictable behavior of every weird new thing we encounter. This can be the behavior of other humans, but also the wasp buzzing around our ice cream cone, the self-driving taxi bearing down on us in the crosswalk, the hailstorm coming over the horizon, and so on. Getting through life is all about the need to predict what is going to happen next, and anthropomorphism is a shortcut to helping us make these predictions.

Psychologists believe that humans are particularly sensitive to stimuli in our environment that indicate the presence of other agents (that is, things that can act with intention and purpose).[21] If we hear a rattling coming from our trash can, we suspect that there is something or someone (i.e., an agent) doing the rattling. And immediately after detecting the presence of that putative

INTRODUCTION

agent (for example, a raccoon), our minds then quickly anthropomorphize it, assuming it has not just intentions but human-like intentions. This process is part of what's called having a theory of mind—that is, theorizing that other creatures have minds full of thoughts, ideas, and beliefs like we have. Using theory of mind, we might guess that the raccoon "wants" to eat the tortellini we tossed in the garbage last night because raccoons "love" tortellini. Guessing what the raccoon "wants" is a helpful (and often correct) tool for predicting the raccoon's behavior, even if we're wrong about the raccoon's true desires and intentions. Anthropomorphism, then, might've evolved as the default mode of thinking, a cognitive bias obliging us to see a world made up of agents (humans, animals, machines, weather, etc.) that desire to effect change in the world (hence the word *effectance*). And not just any agents, but ones with desires that are similar to the desires we've been experiencing since childhood.

If anthropomorphism really did evolve as a prediction-making device to help us guess the desires of other agents (whether those agents have desires or not), then it helps explain why humans are more likely to anthropomorphize when they find themselves in heavily unpredictable situations. Studies show that people are more likely to report that their computer "has a mind of its own" if it crashes at random, unpredictable intervals.[22] I have harbored the not entirely serious but also not *not* serious belief that all the printers of the world hate me. The more temperamental the objects in our lives are—whether that's recalcitrant printers, crashing computers, or cars that won't start—the more likely we are to anthropomorphize them. Anthropomorphism might have evolved as the primary method for our minds to make sense of a scary and unpredictable world by assuming that everything that exists has desires and intentions similar to our own.

As we will see in this book, even when we anthropomorphize things that have no agency at all, there are still benefits. As a psychological predisposition resulting from the forces of natural selection, our capacity and desire to anthropomorphize all non-human things has benefited us as a species more than it has harmed us. "It's a reflection of our brain's greatest ability rather than a sign of our stupidity," is how Nicholas Epley describes it.[23] Epley and his research partners have long had the goal to "move anthropomorphism into the realm of ordinary cognitive processes," as opposed to being some form of human thinking gone wrong, and I want this book to help nudge public opinion on that front.[24] Anthropomorphism, you see, is not only ordinary but massively beneficial. The anthrozoologist James Serpell notes that early human hunters might well have had an advantage over Neanderthal neighbors because anthropomorphism allowed us to make better predictions about the behavior and movements of the animals we were tracking. "Anthropomorphic thinking helped *Homo sapiens* to become a super-predator," argues Serpell, "by providing him with a specialized weapon for penetrating and exposing the minds of his prey."[25]

If It Quacks Like a Human...

Before we get into all the juicy and bizarro stories of anthropomorphism gone wrong, there is one final concept we need to discuss: anthropomorphism triggers. Even though humans evolved the ability to anthropomorphize non-human things with ease, it doesn't just happen automatically. We need triggers to kick-start the process, stimuli that let us know that the thing we are encountering has humanish qualities (and thus might actually be a human). There are three main subconscious triggers that

INTRODUCTION

kick-start our anthropomorphism: eyes, movement, and language. If a non-human thing has humanish eyes or a humanish face, moves in a humanish way or at a humanish speed, or appears to be trying to communicate in a way that suggests that they have the capacity for human language, our brains will be primed to start interacting with that thing as if it were in fact a fellow human. We will see throughout this book how powerful each of these triggers is on its own, and how unstoppable they are when they occur all together. We'll also uncover plenty of secondary triggers, like having human-like hair, odor, or skin, or just generally being a human-like size or having a human-like number of limbs.

The more of these humanish triggers we encounter, the stronger the likelihood that we are, in fact, encountering another human mind. The fact that more triggers generate a stronger feeling of anthropomorphism explains why we consider animals like chimpanzees to be more human-like than sea anemones. Despite a sea anemone being a fellow animal, its plant-like body flopping about in the ocean currents generates almost no anthropomorphism at all. We will see how a lack of anthropomorphic triggers causes us to incorrectly assume the absence of otherwise present human-like cognitive properties, such as problem-solving abilities, emotions, and consciousness, for species that look and act less human-like. This trigger mismatch also explains why we dehumanize other humans who look, think, speak, and move in ways less similar to ourselves or to the people we encounter in our immediate community. We are less likely to attribute human-like qualities, agency, and even humanity to those less similar to us. This is a major contributor to—if not the root cause of—racism, ableism, tribalism, jingoism, and a whole host of other nasty isms. And it's not the case that all humans are equally susceptible to these triggers. Some people are more prone to anthropomorphize than others, as we will learn. And some cultures are more likely to engage in anthropomorphism, which is how *kawaii* (cute) culture came to dominate Japan.

Throughout the book, you will see how these anthropomorphism triggers cause us to treat not just animals but objects, ideas, concepts, weather patterns, software systems, and so forth as if they had human-like minds. We will learn how we have conscious control over our anthropomorphism—allowing us to change the degree to which we anthropomorphize even in the absence of any triggers. We can choose, for example, to pretend that a leaf is feeling sad as the autumn wind blows it from the maple branch. Anthropomorphism is at the heart of so much of our art (the Great Sphinx of Giza), poetry (ravens uttering "Nevermore!"), and literature (the pigs Snowball and Napoleon from *Animal Farm*), as well as the basis for both child and adult pretend play. But it's also at the heart of our ability to dehumanize—to ignore the triggers telling us of the presence of a human mind, allowing us to adopt moral positions that permit us to harm another human.

Because of the power of these triggers to subconsciously force our minds to anthropomorphize, it has long been used as a tool for marketers, politicians, and propagandists to manipulate our behavior. We will learn how anthropomorphism is used by creators—both artists and product designers—to stimulate triggers that give us positive feelings toward non-human things, concepts, and inanimate objects. I'll delve into the world of mascots—those giant stuffed anthropomorphic creatures used to create brand loyalty to sports teams, breakfast cereals, and, in the case of the mascot Jimmy Hattori, condoms. We'll see how much of the research into AI and robotics is about finding ways to design products that trigger us to anthropomorphize so we stop freaking out so much at the coming AI revolution. And we'll hear about

INTRODUCTION

the ways in which anthropomorphism is a fundamental aspect of our religious and spiritual lives. Some academics suggest that the human propensity for believing in the supernatural is itself a psychological offshoot of our capacity for anthropomorphizing. Where once anthropomorphism might've been a label to chastise us for seeing human qualities in God, there is now an argument to be made that God itself is an illusion our minds generate through anthropomorphic thinking.

Fauxnads Redux

We've now come full circle in our crash course in anthropomorphism: an enigmatic circle that dangles before our eyes, clonking with anthropomorphism-edifying promise. Which brings us right back to Gregg Miller's Neuticles. Miller made his millions because he channeled his anthropomorphism into solving a problem for his dog that likely was not actually a problem. By coming to grips with how anthropomorphism so deeply permeates our minds and influences our behavior, we can prevent ourselves from being fooled into thinking that a dog or chatbot or stuffed animal is feeling things that it simply cannot feel. Conversely, knowledge of how to properly wield anthropomorphism can help you correctly determine the presence of humanish qualities in animals and fellow humans where you've been underestimating how closely their mind resembles yours.

I wrote this book because I believe that anthropomorphism gets an unwarranted bad rap. It's still used as a dirty word, especially when it comes to mocking people for anthropomorphizing animals. It has a long and sordid history in the scientific study of animal minds, but it's time to pull it out of the mud, polish it up, and see it for what it is: a misunderstood but utterly charming linchpin of the human mind. I am hoping to convince you that is a boon, not a problem, as far as evolution is concerned. I want you to fall in love with anthropomorphism, to respect how powerful—and how fun—it is, and to help you realize how fundamentally (yet secretly) it has shaped your worldview. And while there are plenty of ways in which anthropomorphism does make us delusional at times, we can mitigate these dangers by learning to be aware of when and how anthropomorphism is tugging at our minds. That awareness can help us use our natural anthropomorphistic tendencies to create healthier relationships with the animals, the AI systems, and even the other humans in our lives. Anthropomorphism can and *should* be a pleasurable activity, and this book will help you learn to enjoy it without any of the negative consequences. Ultimately, I hope you're left with a deep appreciation for the beauty and power of anthropomorphism. And that you won't feel too bad for either Wojtek (who cares too much about his dog's testicles) or his dog Monty (who doesn't care about his testicles at all).

CHAPTER 1

FUR BABIES

Why We Treat Pets like People

Early man domesticated dogs for companionship. And cats for whatever we have cats for.

—*Philomena Cunk*[1]

Two weeks after Abraham Lincoln's second inauguration, he received a telegram from Ulysses S. Grant asking, "Can you not visit City Point for a day or two? I would like very much to see you and I think the rest would do you good."[2] It was March 20, 1865, and Lieutenant General Grant was at a critical moment in the siege of Petersburg. One hundred and twenty-five thousand Union soldiers were lining the trenches between the Confederate capital, Richmond, and the city of Petersburg: locations critical to the supply of General Robert E. Lee's Confederate Army. Grant wanted Lincoln to join him at this pivotal moment in history, where the Union could potentially cripple Lee's forces and bring about an end to the war. Grant could not have predicted that this would also be a pivotal moment in solidifying Abraham Lincoln's reputation as an incomparable cat lover.

HUMANISH

Colonel Horace Porter was in the telegraph office at Grant's headquarters at City Point, Virginia, when Lincoln arrived on March 24. Porter recounted Lincoln's peculiar behavior after he received a telegram detailing the Union's victory at the Battle of Fort Stedman earlier that morning:

Three tiny kittens were crawling about the tent at the time. The mother had died, and the little wanderers were expressing their grief by mewing piteously. Mr. Lincoln picked them up, took them on his lap, stroked their soft fur, and murmured: "Poor little creatures, don't cry; you'll be taken good care of," and turning to Bowers, said: "Colonel, I hope you will see that these poor little motherless waifs are given plenty of milk and treated kindly." Bowers replied: "I will see, Mr. President, that they are taken in charge by the cook of our mess, and are well cared for."[3]

This was not unusual behavior for Abraham Lincoln. He was known to be a lover of all animals, especially cats. When Lincoln first moved to the White House, he was forced to leave his beloved dog Fido back in Illinois because Fido was too anxious to handle the hustle and bustle of Washington. To ease his broken heart, Lincoln was gifted two cats by Secretary of State William Seward: Tabby and Dixie, the first cats to live at the White House. Lincoln doted on them, feeding them at the table during state functions and declaring that "Dixie is smarter than my whole cabinet! And furthermore, she doesn't talk back!"[4]

Two days after finding the kittens in Virginia, on March 26, Lincoln, Grant, and General William T. Sherman held a famous meeting aboard the *River Queen* where they discussed the Union's military plans. It was a defining moment in US history, a meeting that would lead to the victory and subsequent expansion of the Union and the rise of the United States as a military and economic superpower. But Lincoln still couldn't get those kittens out of his head. As Porter recalled, he seemed obsessed with the war and those cute kitties in equal measure:

> Several times during his stay Mr. Lincoln was found fondling these kittens. He would wipe their eyes tenderly with his handkerchief, stroke their smooth coats, and listen to them purring their gratitude to him. It was a curious sight at an army headquarters, upon the eve of a great military crisis in the nation's history, to see the hand which had affixed the signature to the Emancipation Proclamation, and had signed the commissions of all the heroic men who served the cause of the Union, from the general-in-chief to the lowest lieutenant, tenderly caressing three stray kittens. It well illustrated the kindness of the man's disposition, and showed the childlike simplicity which was mingled with the grandeur of his nature.[5]

Lincoln would stay at City Point until the end of the war, departing on April 8—the day before Lee's surrender at Appomattox. During that time, he visited the war front on several occasions, eventually strolling the streets of the captured Confederate capital, Richmond. But he always returned to the tents at City Point, where he could cuddle the orphaned kittens. Less than a week later, Lincoln was assassinated. The rescue of that trio of orphaned kittens became one of his last acts on earth.

Chances are you've met someone equally as obsessed with cats as Abraham Lincoln. The kind of person who would drop everything (including presiding over one of the most important wars in US history) to pet a cat. Maybe you know someone who dresses their cat up in tiny sweaters and takes them for a stroll in a baby carriage. My mother, for example, was a loud and proud cat fancier who rivaled Honest Abe when it came to her fondness for felines. At the peak of her cat lady powers, she had upward of twelve cats in and around our home, each with its own specialized diet and fur-care routine. So I understand this kitty mania thing on a visceral level. To explain what's going on with folks like Abe and my mom, we'll need to tackle two questions in this chapter: (1) What is it about cute cats and dogs that melts people's hearts? (2) Do the cats and dogs that we love so much feel the same about us?

Nobody Puts Baby in a Corner

The anthropomorphism process that leads to our love for cats and dogs starts with our special relationship with human babies. For most people, seeing an infant triggers an impossible-to-ignore caregiver response. We feel an unstoppable urge to pick them up and cuddle them. Even though babies cannot speak and don't move or behave in a way that suggests that they have the kind of agency you'd see in an adult (that is, evidence of self-control or morality), they most certainly display evidence of having internal, mental experience (that is, the ability to feel hunger, pain, desire, joy, etc.). And this triggers our minds to want to treat them as moral patients—beings that we have a moral duty to care for because they are vulnerable. Our minds evolved a special sensitivity to the morphology and behavior of babies that gives most people an irresistible desire to help them. And it's this same sensitivity—this unique form of anthropomorphism triggered by baby morphology—that engages when we look at the adorable faces of kittens and puppies. Our pets have hijacked our baby-sensitivity neurology to elicit that ancient caregiver response, which is what made Abe and my mother so crazy about kittens.

To see how this baby-sensitivity-hijacking works, let's unpack what happened in Abe's brain when he first picked up those orphaned kitties in Grant's tent that day. Within 130 milliseconds of seeing their little faces, the part of his brain associated with reward anticipation sprang into action.[6] Dopamine shot down a highway of neurons in his mesocorticolimbic system that gave him instantaneous sensations of pleasure tinged with a motivation to *do something* about all the cuteness and vulnerability he was looking at. "Almost before you're consciously aware that you're looking at anything at all, you cannot help but feel compelled by that baby"—that's how the neuroscientist Morten Kringelbach explained to the *Washington Post* how bonkersly fast this cute-face response is.[7] Kringelbach's lab was the first to measure the response, noting that our brains don't respond nearly as fast to non-cute or adult faces. It's a response originally designed by natural selection to reward us for interacting with and caring for a newborn baby: a deep, peaceful kind of empathic pleasure. But in Abe Lincoln's case, his brain was responding not to a baby but to a cat. And there's nothing particularly weird about that. One study found that when shown photographs of the faces of human infants, kittens, and puppies, participants rated all three faces as equally attractive, and way more attractive than the faces of full-grown cats and dogs or otherwise cute-looking teddy bears.[8] Our subconscious attraction to cute baby faces is just as strong for kitten and puppy faces.

But what is it exactly about these cute faces that our brains are picking up on? The facial cuteness trigger, it turns out, is a rather mundane-sounding set of mathematical properties of a face as it pertains to the relative size, shape, and distance between facial features. Way back in 1943 the ethologist Konrad Lorenz proposed the idea of a *Kindchenschema* (translated as "baby schema," "infant schema," or "child schema"), a set of facial characteristics that evolved to trigger our parental, caregiving instincts.[9] In the decades since the *Kindchenschema* idea was proposed, tons of experimental research has shown how tiny adjustments to the features of a face increases our experience of cuteness. Move the eyes a millimeter closer to the midline of the face, which elongates the forehead, and *boom*, cuter. Make the nose a touch smaller or increase the diameter of the eyes just a smidge, and *boom*, cuter. This lightning-fast, subconscious assessment of cuteness develops early: Infants as young as five months will look longer at images of cute babies as opposed to ugly ones.[10] And this perceived cuteness really does alter our behavior: Cuter babies are more likely to be adopted, and we treat adults with baby-like features with more empathy.[11] In one study, manipulating the face of famous politicians to make their features subtly more baby-like resulted in people viewing them as more honest, attractive, and compassionate.[12]

These *Kindchenschema* cuteness calculations our brains are making apply equally to all animal faces, not just mammals or species that look particularly human-like.[13] Show someone an image of an Australian brush turkey chick with its big, adorable eyes and floppy feet and *boom*, cuteness. An axolotl with its goofy oversized head and smiley, toothless mouth? *Boom*, cuteness. Animals that require massive amounts of parental care to survive (known as altricial species) have stronger *Kindchenschema* cuteness features than those species that require less or even no parental care (precocial species).[14] This suggests that there might be a nigh-universal set of features in the animal kingdom associated with cuteness (and thus neediness) that evolved to manipulate adults into caring for their offspring.[15] We humans are at the mercy of a millions-of-years-old "sensory trap" that makes us weak in the knees for innocent-looking baby animals no matter what species we're looking at.[16]

To show how powerful this cute-baby-animal effect can be, consider the actor Kristen Bell's reaction to the *possibility* that she would get to see a sloth. By her own admission, Bell has been "obsessed with sloths" her entire life.[17] On her thirty-first birthday, her boyfriend (now husband), Dax Shepard, arranged for a sloth to be brought to their home as a surprise birthday gift. When the sloth handler knocked at their door, Bell was told that it was her birthday surprise, but not that it was a sloth. She correctly guessed, however, that there was a sloth behind the door, and was so overcome with the idea of meeting a sloth face-to-face that she couldn't bring herself to open the door, and instead retreated to the bedroom weeping uncontrollably with joy. Dax recorded a video of Bell's sloth meltdown, where she is incapacitated on the bed with tears streaming down her cheeks sobbing that she is "so excited" to meet a sloth. Like kittens or puppies, sloths have faces that check every box on the *Kindchenschema* cuteness checklist, and are thus primed to hijack our baby-sensitivity neurology. They even have mouths that are frozen in an adorable little smile. For Bell, the mere *idea* of looking at a sloth's sweet little face literally incapacitated her.

As adorable as this all sounds, *Kindchenschema* originator Konrad Lorenz considered all this blubbering over cute animals to be a huge problem. According to the cuteness scholar Joshua Paul Dale, Lorenz believed that "our propensity to regard anything other than our own babies as cute represented a weakening of our parental instincts."[18] Lorenz argued that it was the "Aryan race" that was the least "domesticated" and thus able to fight against the *Kindchenschema* triggers that generate our dopey responses to cute animals. Yes, that's right: Lorenz was a Nazi, and his ideas about cute kitties or adorable puppies were somehow used to support his eugenicist race-science bullshit.

But don't let the Nazis spoil things. There is a ton of non-racism-motivated science to show that faces optimized to fit the *Kindchenschema* really do trigger us to feel affection for cute things. As we learned in the Introduction, eyes and faces are known to be one of the top three triggers for anthropomorphism, as we will see throughout this book. And *Kindchenschema* faces with their wide baby eyes are a great example. But it's not just cute faces that do the triggering. There are a slew of lesser cuteness triggers in the mix that nudge us to anthropomorphize our pets. Aside from their facial configuration, there are other potential triggers for cuteness, like soft skin, oversized heads, and rounded limbs.[19] These are all things we associate with babies and toddlers that are mirrored in the adorable morphology of floppy-pawed kittens and puppies.

Even the sounds that our pets make remind us of babies. Sounds, it turns out, are cuteness triggers in their own right. Cognitive scientist David Huron asked his students to rate the cuteness of random sounds he played for them, including the sounds of music boxes, animal calls, electrical appliances, closing doors, and other things. He found that "listeners were unanimous in distinguishing those sounds that are 'cute' and those sounds that are 'not cute.'"[20] The sounds that were the cutest were all high-pitched and relatively quiet. The squeaky sounds (the cutest of all sounds) that people liked the most came from toys that were around 20 milliliters in volume. That's about the same volume as the vocal tract of a human baby. In other words, we are attracted to the adorable sounds of babies (their laughter, cooing, and gentle cries) in much the same way we are to *Kindchenschema* faces. Huron called this "auditory cuteness."

Kittens are rife with auditory cuteness, from their high-pitched meows to their soft purring. Remember when Abe Lincoln tenderly wiped the eyes of those Civil War kittens as he listened to them "purring their gratitude"? Well, those kittens' purrs were likely what scientists call a solicitation purr. Unlike the standard purring that happens when a cat is content or happy, a solicitation purr is used by cats who are trying to get us to do something, like feed them, let them outside, or pet them. A solicitation purr has an extra bit of energy found at about 490 Hz, giving it a subtle high-pitched flavor. That's right smack in the middle of the frequency range of a human infant cry, which is between 300 and 600 Hz.[21] More than likely, the domestic cat's solicitation purr is a huge trigger for auditory cuteness—forcing us to respond to our cats like we would a cooing baby.

Bred to Melt Hearts

Now that you know that animal cuteness has a direct, subconscious line to our baby-loving brains, it makes more sense why the internet is filled with pics of adorable animals like pandas, otters, or the other species that have *Kindchenschema* features. But there's another piece of the puzzle we need to add when it comes to explaining the absolute dominance of cats and dogs on the internet, in our homes, and in our hearts. Our prehistoric human ancestors spent thousands of years breeding cats and dogs to make them even more irresistibly cute than their wild progenitors.

For example, in the ten-thousand-ish years since cats started living with us, they have gotten quantifiably cuter. Aside from being smaller, domestic cats have significantly shorter noses than the current-day African wild cats from which they evolved.[22] In recent years, cat owners have become obsessed with small-nosed cats, and the most extreme version—the flat-faced breeds like the British shorthair or the Persian—are in high demand. These brachycephalic breeds (the term is derived from the ancient Greek *brakhús*, meaning "short," and *kephalé*, meaning "head") are part of what veterinarians are calling the "brachy boom."[23] Unfortunately, these breeds—which were created through human-driven selective breeding—are riddled with health problems. The main glitch is that while these breeds' skulls were bred to be smaller, the soft tissue in their heads didn't shrink. So all that nasopharyngeal tissue crammed into a tiny skull means that their airflow is restricted. In the worst cases, cats can develop brachycephalic obstructive airway syndrome—a chronic condition that can lead to laryngeal or bronchial collapse, aspiration pneumonia, obesity due to an inability to exercise, gastroesophageal reflux, heat intolerance, and generally just having a shitty life because the cat can't breathe properly.

We've bred the same problems into brachycephalic dog breeds like Boston terriers, bulldogs, and pugs. I've lived with a Boston terrier before and can attest to the fact that they are both adorable and a health nightmare. According to the anthrozoologist James Serpell, bulldogs have it even worse. "Perhaps the most extreme example of this process can be found in the English Bulldog," wrote Serpell, "with its severely brachycephalic head, prognathous upcurved mandible, distorted ears and tail and ungainly movements. Once a powerful, athletic animal, and now...the canine equivalent of a train wreck."[24]

It's possible that we subconsciously enjoy how sickly these brachycephalic breeds are. Pets with glaring health problems require constant care and attention, emphasizing their status as moral patients. Tiny, constantly shaking Chihuahuas with their panicked-looking, bulbous eyes and fragile, spindly legs radiate a "please help me" vibe that we can't seem to resist. "Thus," argues Serpell, "it could be argued that humans have selected unconsciously for small, anxious, needy, unhealthy and vulnerable companion animals—animals with inherently compromised welfare—because these are precisely the traits that best satisfy their desire for things to nurture and parent."[25]

Maybe it's a stretch to think that we have Munchausen-syndromed our Chihuahuas, but it is undeniable that brachycephalic pets possess facial features that conform to the cuteness schema and that this will 100 percent trigger our baby-face-loving anthropomorphism. When scientists analyzed the facial morphology of brachycephalic dog breeds, they found that they "show exaggeration of some, but not all, known fronto-facial 'kindchenschema' features, and this may well contribute to their apparently cute appearance and to their current popularity as companion animals."[26] So even though our ancestors might not have been aware of *Kindchenschema* as a psychological concept, it's clear that it influenced their breeding preferences.

Eyebrow-Raising Cuteness

But we didn't just breed dogs to look cuter. We bred them to *act* cuter. Remember how movement was one of the top three anthropomorphism triggers? Well, dogs evolved a very specific human-ish movement that combines with the presence of eyes, sending our anthropomorphism into overdrive. I'm talking about those sad doggy eyebrows.

Domestic dogs have muscles that can control their eyebrows, but their ancestors—wolves—do not. There is a set of muscle fibers above a dog's eye called the levator anguli oculi medialis (LAOM) that allows them to raise the skin above the eye and make their eyes open wider. A second set of muscles at the outer corner of the eye—the retractor anguli oculi lateralis (RAOL)—is found in most domestic breeds, with the exception of the wolf-like Siberian husky. These muscles allow dogs to execute an "inner eyebrow raise" movement that as the authors of one recent study on this subject noted, is "particularly attractive to humans."[27] This movement "causes the eyes of the dogs to appear larger, giving the face a more paedomorphic, infant-like appearance, and also resembles a movement that humans produce when they are sad." This puppy-dog-eye behavior pulls at human heartstrings, eliciting an even stronger caregiver response than a static childlike face.

What's more, domestic dogs—unlike wolves—regularly maintain eye contact with their human companions, which research has shown sparks the release of oxytocin in both humans and dogs. Given that "oxytocin plays a primary role in regulating social bonding between mother and infants," it's no wonder that our hearts melt when our dogs gaze up at us with those big puppy eyes.[28] In the roughly thirty-three thousand years since dogs were domesticated, we managed to breed into them sad-looking eyebrows the sight of which causes the release of oxytocin in us, increasing the power of the *Kindchenschema* to an absurd degree.

Eye and eyebrow cuteness doesn't just spark an oxytocin-driven caregiver response that makes us want to treat our pets as emotion-laden moral patients; it is also the catalyst for a far more powerful, social response stemming from our belief that our pets have human-like mental experiences. According to Wegner and Gray, eyes and eyebrows contain information about both agency and experience: "Agency-wise, eyes can convey focus of attention, which suggests intention and the next likely course of action;

experience-wise, eyes can convey emotion by narrowing in anger or widening in fear."[29]

When we see a pair of adorable, expressive doggy or baby eyebrows, we are driven not just to care for the dog or baby but to interact with it, on the underlying assumption that it wants to interact with us. These responses are predicated on the unspoken idea that the thing we are interacting with—be it a baby or a puppy—has a mind like ours that we can connect with. A mind that not just experiences emotions but has the capacity to communicate to us about those emotions. As Morten Kringelbach and colleagues described it in their review of the science of cuteness, "Cuteness may even go beyond eliciting caregiving to facilitate complex social relations by triggering empathy and compassion. Cuteness is a general promoter of sociality acting through mentalisation, the ability to treat infants and even inanimate objects as psychological agents."[30]

This brings us to the second question we'll be tackling in this chapter when it comes to the fur baby phenomenon: Do the cats and dogs that we love so much feel the same about us? We're driven to interact with them socially because of those expressive eyebrows on solicitation purrs, but how do *they* feel about all this interaction?

When we anthropomorphize our pets, we can't help but imagine, pretend, or maybe even believe that they have minds filled with thoughts and emotions similar to our own. When I asked my friends for examples of how they engage in anthropomorphism, my friend Dave said that he has "a child-like dog voice that I use to narrate my dog's supposed thoughts as she moves around the world doing dog things that I imagine have complex human motivations." Similarly, my friend Diane has always narrated her cats' thoughts. "I get quite attached to the whole voice and character I have imposed upon these cats, and when a beloved cat dies, that voice retires," she explained.

Thought narration is common among pet owners and seems to crop up automatically when we interact with our pets, "most people aren't even aware that they're doing it," explained dog cognition expert Alexandra Horowitz to the *Washington Post*, "but are kind of unconsciously bringing the dog into the human conversation."[31] But it's hard to know if our pets really have these thoughts, or if we're unjustly projecting our own ideas onto them. At least, it was hard... until dog buttons came along.

Canine Existentialism

To explain how dog buttons became internet evidence that dogs (and cats) have complex internal thoughts, I need to introduce you to Bunny the dog. Born on July 8, 2019, Bunny is a sheepadoodle (three-quarters poodle and one-quarter Old English sheepdog) that lives in Tacoma, Washington, with her "mom," Alexis Devine. Bunny has many millions of TikTok followers who tune in to watch her use her button board, a collection of more than a hundred plastic buttons that Bunny can press with her paw.[32] When pressed, the buttons play an audio recording, allowing her to "talk" with Devine by stringing together a series of words. There are buttons for simple nouns like *poop*, but also complex concepts related to the passage of time like *later* and *afternoon*. There are verbs like *smell, feed, do,* and *talk*; adjectives like *good* and *bad*; pronouns like *I, you,* and *we*; interrogative words like *who, what, where, when,* and *why*; words related to emotions like *concerned, happy,* and *mad*; and abstract concepts like *stranger, person, friend, dream,* and *medicine*.

Videos typically involve Bunny stepping on a series of buttons and looking to Devine, who tries to puzzle out what Bunny is trying to say. Most of the time it's a straightforward request: *outside*, *poop*, or *play*. But occasionally Bunny will string together a series of words that result in what her online followers consider evidence of Bunny having an "existential crisis." In one video, while looking in a mirror, Bunny steps on the buttons to make the phrase "Who this?" She then goes to stare wistfully out the window for a moment before stepping on *help*. If this video were in black and white, you'd be forgiven for thinking it had been directed by Ingmar Bergman. In other videos, she appears to ask deeply philosophical questions like "Why is?," "What Bunny?," and "Dog what dog is?" Wasn't it Jean-Paul Sartre who said, "Consciousness is a being such that in its being, its being is in question insofar as this being implies a being other than itself"?[33] Sartre and Bunny seem to be channeling the same grammatically ambiguous, existential energy.[34]

So what is Bunny really thinking? Does she understand what these words actually mean? Sarah-Elizabeth Byosiere, director of Canine Research and Development at Guide Dogs for the Blind, told the *Washington Post* that "the short videos I see online seem to indicate that dogs are able to form associations between a button press and an outcome, but it's really difficult to say if anything more is happening." Alexandra Horowitz of Barnard College's Dog Cognition Lab explained at a TED event that "there's no evidence that [button apparatuses] actually represent what the dogs are thinking or are extending their communication ability" and that these devices "have yet to be completely subject to scientists' scrutiny."[35]

Clive Wynne, founding director of the Canine Science Collaboratory at Arizona State University, is skeptical that dogs like

Bunny understand the meaning of the words on their button board. "I think that all of the meaning is being brought in by the human being who is listening," Wynne told the *Washington Post*. "Random button presses that we, the human being observing the dog, breathe the magic breath of language into."[36]

I visited with the animal communication scientist and author Arik Kershenbaum, author of the book *Why Animals Talk*, to get his take on this talking-dog phenomenon. "You can train a dog to come when you say 'Come,' and when you say 'Do you wanna go for a walk?' the dog understands and wags his tail and looks excited," Arik told me. "There's definitely communication going on there, and it's more than that: It's referential communication. Is it language? No, of course not. They have learned to associate sounds with concepts. Could they do that with buttons? Of course."

For animal communication nerds like Arik and me, the word *language* has a very specific meaning. It refers to the complex syntactic system humans use to combine units of meaning to represent any and all thoughts we have in our heads. It does more than just communicate our emotional state or intentions (which is what animal communication systems are mostly focused on). Instead, language can capture and convey the meaning of concepts like "capitalism" or "epistemology" or "Swiftie"—concepts that are not just abstract but likely to be distinctly human.

So what does the science say about dogs' language abilities? Research into dogs' understanding of human speech shows that they are unusually skilled at learning to match words to objects (like their toys) and simple concepts like "walk" or "outside," but there's no evidence that they understand abstract concepts like "tomorrow," "stranger," or "medicine." In a landmark study published in 2011, the behavioral psychologist John W. Pilley outlined how his research team trained Chaser the border collie to learn the names of 1,022 different objects. The objects consisted of "over 800 cloth animals, 116 balls, 26 'Frisbees,' and over 100 plastic items. There were no duplicates."[37] In the entire history of humankind, there has never been a non-human animal that has learned as many words as Chaser. None of the famous language-trained animals like Koko the gorilla, Akeakamai the dolphin, or Alex the parrot has learned anywhere near the number of words as Chaser. Each of these 1,022 words, however, referred to a real object, not an abstract concept.

Dogs are also remarkably good at processing human speech sounds. They can tell the difference between regular human speech and fake speech that doesn't use any real language.[38] Their brains seem wired to pick up on natural human speech sounds, as opposed to any old sound coming from a human mouth. Other research shows that dogs' brains are more sensitive to words they hear more frequently from their owners, as opposed to nonsense or new words. But this doesn't necessarily mean that they have a human-like capacity to parse words. Unlike humans (or some border collies), most dogs can't learn to differentiate between similar-sounding words, like *slipper, Swiffer, sniffer, sipper, supper,* and *slapper*.[39] In another experiment, researchers scanned the brains of dogs while they listened to their owners using that singsongy baby-talk voice (called infant-directed speech) that people use with both their pets and their babies. This "who's a good boy" way of talking to dogs involves more variation in pitch, shorter phrases, and higher frequencies than normal speech. And it's far more prevalent in human women, who have been shown to "hyperarticulate" their vowels and "use a wider pitch range" than men when talking to babies and pets.[40] Human babies' brains evolved to be attuned to this kind of speech, which seems to hold their attention more than normal speech. It turns out that dogs' brains also show "enhanced neural sensitivity" to this singsongy way of speaking. This is an intriguing finding since these kinds of sounds are not "typically used in dog-dog communication," explained Anna Gábor, an author of the study.[41] Instead, dogs evolved a response to human infant-directed speech during the domestication process.

As much as it's clear that dogs really do have a special affinity for human speech, none of the research shows that dogs are capable of understanding abstract concepts that make language what it is. So skeptics, like the biologist and social media influencer/educator Kristyn Plancarte, have good reason to give dog button videos the side-eye. "My observation is that the dogs have learned to push buttons in order to get the attention that they desire from their owners," Plancarte explains while reacting to a video of Bunny. "And it doesn't really matter what button it is. They get the same amount of attention for all of them. 'Stranger'? 'Friend'? 'Person'? Dogs don't have these concepts."[42]

This is a possibility that Bunny's owner herself has considered when thinking about her sheepadoodle's behavior. "I do have to say, 'OK, how much am I reading into this?'" Devine told the *Washington Post*. "How much of this is anthropomorphized?"[43] In an interview with *Vice* she said, "It's really an emotional roller coaster for me.... There are some days where I am so frustrated and I believe it's all random and the skepticism in me overrides everything else. Then there are days when almost every single utterance is clean and concise and contextually appropriate and I'm like, 'There's definitely something going on here and it's not random.'"[44]

Even if what Bunny is doing is pressing buttons randomly to elicit attention from Devine, you cannot fault Devine or any other dog button users for defaulting to anthropomorphism. Language is an irresistible anthropomorphism trigger. And there's really no harm in letting anthropomorphism wash over us as we interact with our button-pressing dogs. Even the most skeptical of scientists agrees that dog buttons are a force for good, regardless of what the dogs are really thinking.

The False Mirror Conundrum

So if dog buttons are not the smoking gun that proves that our pets have complex internal thoughts, is it possible that they're just another example of us being bamboozled by anthropomorphism? Remember that the term *anthropomorphism* has historically been used to chastise people for treating animals as if they had human-like thoughts when they do not. And for some people, the answer to the question "Do the cats and dogs we love so much feel the same about us?" is easy: No, because not only do they not have complex internal thoughts, but they have no capacity to love or even feel anything at all.

As absurd as this stance sounds to anyone who has spent time with a cat or dog, it was only a few decades ago that most scientists believed that animals didn't have any internal, conscious experiences or understanding of the world. So Abe Lincoln anthropomorphizing his kitty cats would have been considered fully irrational, always weird behavior. In the next chapter we'll learn more about how scientists justified this stance. But for now, let's see how expressive doggy eyebrows—a behavior that seems to be a direct reflection of a dog's mental or emotional state—could have evolved in a hypothetical world where dogs feel nothing at all. Now, I personally don't believe this is the case, but this little exercise will help us understand how there could be a fundamental mismatch between the triggers that generate our anthropomorphism and the reality of what's happening in the minds of our pets.

Imagine a group of early humans first interacting with a group of wolves thirty-three thousand years ago. Maybe the wolves had been hanging around these humans' evening fires for a few years and both species had gotten comfortable with each other. Among these friendly wolves is a wolf puppy with floppy paws, an adorable high-pitched whine, and large *Kindchenschema* eyes. Inevitably, these humans would have found this puppy to be super cute. But what if that little wolf pup also had an unusual set of eyebrows that flexed ever so slightly, making her eyes that much bigger and more adorable? Before too long, those early humans might have taken that pup under their wing and bred her with other eyebrow-flexing wolves until one day the first fully eyebrowed domestic dog appeared. Now, for the sake of argument, imagine that those early dogs had no capacity for thinking or feeling. They could experience no pleasure or sadness—they were just little biological robots with nothing resembling human cognition or a capacity to consciously experience emotion. Those eyebrow movements would be no indication of their inner lives at all. It was the humans who gave them expressive eyebrows through selective breeding because they liked the look of them. But the dogs had nothing to express. Anthropomorphism would've coerced us to give dogs eyebrows to satisfy our love of cuteness. We would see ourselves reflected in their otherwise vacant eyes—like looking in a false mirror, as Francis Bacon said. This hypothetical scenario can explain how anthropomorphism triggers can delude us into believing that animals possess cognitive skills that they otherwise lack.

If this scenario sounds like an absurd thought experiment, consider that when I was born in the 1970s, most pediatricians were convinced that newborn babies could not experience pain. That if you pinched a baby and it cried, this was a kind of automatic response generated by the brain that didn't manifest as the conscious experience of pain. A quick look through the scientific literature from that period shows that "studies of neurologic development concluded that neonatal responses to painful stimuli were decorticate in nature and that perception or localization of pain was not present."[45] It wasn't until the 1980s that a new wave of research "seemed to overcome the denial of infant pain that was entrenched in medical practice and experimental science for over a century, thus opening the door to infant pain management," as one 2020 review of the subject found.[46]

My friend Genevieve is a nurse who worked in a neonatal intensive care unit (NICU) in the 1990s and confirmed that the doctors she worked with at the time were taught that babies under the age of one could not experience pain. "This is why they never gave pain meds for circumcisions or for medical procedures in the NICU," she explained. "They figured the pain receptor system wasn't fully formed." But Genevieve never believed that. "The moment I held my first baby during a circumcision with no meds, I knew for sure they felt pain," she said. These days, science has concluded that babies *do* experience pain, and, thankfully, pain medication is now generally administered during neonatal surgical procedures like circumcisions. But thirty years ago, doctors would have mocked Genevieve for arguing that the crying baby in her arms with a mangled penis was suffering. It would have been a case of an overly empathic nurse unjustly projecting adult human–like feelings onto a neonate in the same way Abe Lincoln unjustifiedly projected human-like feelings onto cats.

In most cases where we are wrong about what's going on inside an animal's mind, it's not because we are projecting human-like thoughts or emotions onto an animal that is otherwise devoid of any capacity to think and feel. This is an extreme example, although, as we will soon learn, it was a commonly held belief not that long ago. Instead, it's that anthropomorphism triggers are making us treat an animal as if it had human-like thoughts or emotions when they're actually experiencing a completely *different* set of thoughts and emotions. The philosopher Mike Dacey's example of this phenomenon is the story of Ham the chimpanzee, who became the first great ape to orbit Earth in 1961.[47] In a famous picture of Ham taken as he was being transported to the rocket ship before his launch, he can be seen grinning widely while strapped securely into his flight capsule. He looks, to the untrained observer, excited, happy, or possibly even proud. But chimpanzee experts immediately recognize the meaning of Ham's facial expression: abject fear. Chimpanzees smile widely like this in what is called a fear grin, displaying their teeth in an expression that bears an uncanny resemblance to a human smile. Even if we intellectually understand that Ham is scared in that picture, our anthropomorphism triggers are subconsciously nudging us to see joy in that smile.

So the answer to the question "Do the cats and dogs we love so much feel the same about us?" is: possibly. But there's also a chance that we're misunderstanding the exact nature of their thoughts because of the anthropomorphism and cuteness triggers that we've bred into them. We might be misinterpreting their communicative signals. Do you know that "guilty look" that dogs give you (raised eyebrows, ears back, tail between their legs) after they've done something that's not allowed (like eaten something from the table) and you scold them? Well, the science is quite clear that that look isn't necessarily guilt: It's a look of appeasement or submission that occurs whenever you scold a dog.[48] "They do it if we just give them an angry face," explained Alexandra Horowitz (who first investigated this phenomenon) to the BBC, "even when they haven't done anything wrong. That's a real sign that it's not because they're feeling guilty."[49] Just like the lesson from Ham the chimpanzee, the guilty-dog phenomenon is another example of the false mirror: We can't help but see our minds reflected in the faces of other species.

How Many Licks Does It Take?

I want to close this chapter with one final story of the potential pitfalls of the false mirror, this kitten-faced, sad-eyebrowed cuteness, or button-pressing talent that is potentially blinding us to the reality of our cats' and dogs' true inner thoughts. My wife and I joke that our cat Oscar would more than likely kill and eat us if he were larger. He's adorable and affectionate, but he has these sudden spells where he attacks our feet—usually if we're wearing a new pair of socks. He'll sink his teeth right into our ankles. If he had jaws the size of, say, a golden retriever, we'd not likely survive this kind of mini-mauling. But the reality of this situation is no joke. Domestic cats have been shown to have no qualms about eating their owners.

In one gruesome case, a sixty-nine-year-old man in Australia died in his home from unknown causes—likely from complications of either diabetes or epilepsy.[50] The medical examiners were unable to determine the cause of death, however, because his corpse had been scavenged by the thirty cats living in his house.

When police officers entered his home, they found his body on the floor next to his couch "with his face gnawed down to the skull and his heart and lungs gone," as Sara Reardon described it for *Science* magazine. "As if to dispel any doubt about what happened, one cat was still sitting inside the man's emptied chest cavity."[51]

The cats had been snacking on the man's corpse for a few days. This is, apparently, not uncommon behavior when cats encounter a dead human, as the unexpectedly abundant scientific literature on postmortem pet predation makes clear. Dogs too have been known to eat the bodies of their dead owners. The fact that pets enjoy (or at least have no misgivings about) eating their owners doesn't necessarily mean that they don't feel human-like affection for us. But it does highlight the un-human-like way that cats and dogs experience the world, and how dissimilar their cognition is likely to be when it comes to the social norms around eating your dead friends. A dog might very well experience sadness when their owner dies. But this doesn't stop them from eating us. The dog might, as Reardon describes, "lick their owner's face seeking comfort," but "that licking can quickly turn into feeding." As forensic anthropologist Carolyn Rando put it: "I think we have to come to the conclusion that our pets will eat us. It's just a fact of life."

The takeaway from this chapter is that our brains are wired to make us respond to our pets as if they were helpless babies and want to interact with them as if they had human-like minds. It's caused by several anthropomorphism triggers bubbling away in our subconscious that we might not be aware of and cannot easily suppress. Luckily, even if this kind of anthropomorphism makes us wrong about what's going through our pets' minds some of the time, it usually increases the pleasure we get from interacting with out pets, and often (but not always) results in our pets leading happier and healthier lives. This is all well and good, but it's admittedly kind of annoying to have to accept that anthropomorphism might be making us wrong about what's going on in our pets' minds, even if it's the basis for a beneficial relationship. The question we'll be tackling in the next chapter is: How can science help us figure out what cats, dogs, and other animals are *really* thinking? Were Abe Lincoln's kittens just biding their time, tolerating his affection, so that they could eat his face when he died? How can scientists figure this out without letting all this eyebrow-raising, button-pressing *Kindchenschema* cuteness bamboozle us?

CHAPTER 2

DOLPHIN DOULAS

How Anthropomorphism Makes Scientists Crazy

> The talking animals on television, the satirical depiction of public figures, and the naïve attribution of human qualities to animals have little to do with what we know about the animals themselves.
>
> —*Frans de Waal*[1]

In December 2011, I received a pleasant email from a young couple seeking advice on the best place to give birth with wild dolphins. They had purchased a custom boat with a trampoline-like platform that could be raised and lowered into the water. During labor, the soon-to-be mom could hang out on the trampoline while submerged a few feet in the ocean. This would give her a surface so she could settle herself in a comfortable laboring position but still allow for wild "dolphin doulas" to swim up to her and "help" with the birthing process.

The couple were hoping there was a good location in the Bahamas where this would be possible and had contacted my dolphin research organization (the Dolphin Communication

Project) looking for thoughts and feedback. This was not their first attempt at giving birth with wild dolphins; they had traveled to the Black Sea a few years earlier hoping to give birth alongside the bottlenose dolphins that live there, but ended up instead having the baby in a shallow, secluded tide pool off the coast of Spain. No dolphins present—probably just some hermit crabs and limpets and whatnot. I've seen the video of that birth and can confirm that (1) it really did happen and (2) it was not as weird as it sounds.

I had a productive exchange with this couple, but I am going to wait to tell you what advice I offered them until the end of this chapter. I need you to first understand how we have arrived at this situation, where some people have anthropomorphized dolphins to the point where they believe that dolphins are genuinely interested in helping humans deliver their babies. Understanding the origins of dolphin midwifery and why it seems plausible to some people is the key to understanding why researchers fight over the use of anthropomorphism in science, and how getting it wrong can blind us to what's actually going on inside the minds of dolphins and other animals. Spoiler: It's not always the hippy-dippy folks who are wrong—skeptical scientists have a long history of being pretty dippy themselves when it comes to underestimating the cognitive abilities of non-human animals thanks to their distaste for anthropomorphism.

Dolphin Milk

To bring you up to speed on the history of dolphin doulas, here's the abridged, cocktail-party version. Igor Charkovsky was a Russian health guru who came to prominence in the 1970s with his water-based training methods for creating Soviet superbabies. Charkovsky claimed that exposing both pregnant mothers and newborn babies to an underwater environment would result in the children growing up to be "more intelligent and better adjusted than other children."[2] He had a cult-like following of expectant mothers who would train with him in the Black Sea—swimming in ice-cold waters to develop strength and cold stamina, and holding their breath under water as long as possible with the ultimate goal of giving birth while being fully submerged. Water births in the open sea were encouraged, and Charkovsky perfected a technique of keeping a newborn baby and the placenta submerged underwater for up to fifteen minutes before allowing the newborn to take its first breath at the surface. Once born, the babies were put through rather grueling physical training—held underwater for extended periods of time or swung around by their limbs. I saw a video of Charkovsky burying a naked baby in snow before flinging the baby into the air. It was all part of Charkovsky's method for creating superbabies, introduced to the West in the 1982 book *Water Babies* by Swedish journalist Erik Sidenbladh.

According to Charkovsky, it wasn't just water that was needed to make strong and intelligent babies. You also needed dolphins in the mix. Through interactions with dolphins during pregnancy, birth, and baby training, the babies born of a "dolphin-assisted" birth would develop supernatural abilities, including "clairvoyance, telepathy and psychic healing." In a purported experiment from 1979, Charkovsky claimed that "babies swam, dived and even slept in the water near the dolphins" and that in the "very near future a newborn child would be able to live in the ocean with a pod of dolphins and feed on dolphin milk."[3]

Unfortunately, wild dolphins aren't always easy to find in the Black Sea. Luckily, Charkovsky, who was a former Soviet naval officer, had a line on some captive dolphins that would make dolphin-assisted birth and subsequent baby training a lot easier.

His followers would sneak into the Sevastopol Naval Base at night while the dolphin researchers slept. They would attach "different types of saddles and handles" to the military-trained dolphins and encourage their newborns to grab on and go for a ride, which, according to Charkovsky, the dolphins enjoyed. They "willingly allowed the children to ride on them," he claimed.[4]

Dolphins were an integral component of Charkovsky's aquaculture superbaby training method. They could supposedly teach the children to swim faster, provide pain relief for mothers during childbirth via "underwater sonar messages of support," and help the babies reveal their "inner vision" and "third eye."[5] The ultimate goal was to spark a biological and genetic transformation in these babies that would lead to a new super-race of humans with the ability to live in the sea alongside (and apparently breastfeed from?) their dolphin mentors.

To the average Soviet citizen, this wasn't a particularly far-out idea. "In the late Soviet popular science literature and cinema, dolphins were presented as possessors of extra-human intelligence, primordial wisdom and an intimate connection with the universe," suggests the anthropologist Anna Ozhiganova. In researching the origins of Charkovsky's ideas about dolphins, Ozhiganova found that "John Lilly's book *Man and Dolphin*, published in Russian in 1965, likely also served as a source of inspiration for Charkovsky."[6]

Lilly was the famous founder of the New Age dolphin movement. He claimed that thanks to their enormous brains, dolphins had superhuman intelligence, language abilities, and a special affinity for humans. Lilly traveled to the Soviet Union to meet Charkovsky in 1998. After watching him put a few superbabies through their paces, Lilly said he was "convinced that the children Charkovsky worked with had established exceptional contact with dolphins...they understand each other through intuition. Charkovsky's disciples already know how to sleep in the water and find food in the sea."[7]

Dolphin-assisted birth, as popularized by Charkovsky, was deeply intertwined with the idea that dolphins are superintelligent, semi-mystical animals. "Whales and dolphins have a much better use of their brains, on levels unreachable for humans," argued Elena Tonetti, one of Charkovsky's early followers.[8] "Dolphins are more 'human' than many of us," opined Michael Hyson, formerly a researcher with the now defunct Sirius Institute in Hawaii, which once offered dolphin-assisted births.[9] "They are empathetic, telepathic, and often willing to help humans heal."[10]

On the backs of these extraordinary claims, and amid the dolphin mania that had been bubbling away in popular culture since the 1960s, interest in dolphin-assisted birth blossomed in Europe and North America in the 1980s. At the same time, a less fanciful version of Charkovksy's water-birthing idea started to become mainstream. By the mid-1990s, the practice of giving birth in water—usually in indoor birthing pools—had become widely accepted as a legitimate and medically sound birthing technique. But the practice of water birthing had lost its association with dolphins, and modern practitioners were keen to distance themselves from Charkovsky. Sheila Kitzinger, one of the organizers of a water-birth conference in England in 1995, had no time for Charkovksy and his dolphin superbaby methods, telling *The New York Times* that "his movement is a cult."[11]

In the 2000s, dolphin-assisted birth continued to grow in popularity, with many unaware of its original association with Charkovsky and his Soviet superbabies. In 2005, an aquarium in Peru was offering pregnant women the opportunity to have captive dolphins nuzzle their pregnant bellies while making sounds that would "stimulate the nerves in the brain and the child's audible [sic] senses."[12] In that same year, UK dolphin swim tour operators were seeing a huge increase in pregnant women wanting to swim with dolphins. "This year almost a fifth of our dolphin swimmers have been pregnant women," one tour operator told *The Telegraph*.[13] By the time the nice pregnant couple emailed me in 2011, interest in dolphin-assisted births was still at a fever pitch. At that time, the Sirius Institute claimed it was fielding "3–4 requests per week through the Internet from people searching for a place to birth with dolphins."[14]

In the 2014 Channel 4 documentary *Extraordinary Births*, dolphin-birth aficionado Kim Nelli explains how swimming with spinner dolphins in Hawaii positively impacted her pregnancy. She recalls how the dolphins swam circles around her, and says that this caused her breech baby to turn in her belly—supposedly an intentional act by the dolphins as part of their midwifery services.[15] In a statement very much reminiscent of Charkovsky's original ideas, she explained that "children who have been born after having been exposed to dolphins—well, clicks and whistles come out of them...like before they can speak they are making clicks and whistles which are dolphin noises. And they definitely have a lot of characteristics of the dolphins. My daughter is super intelligent, she's extremely creative with art. She's telepathic. She's able to communicate with the dolphins. She tells me messages that the dolphins gave her. It's like she is a dolphin. But I cannot convince anyone to believe my belief systems. Because there's no scientific evidence."[16]

Ah yes, that pesky scientific evidence.

This dolphin-assisted birth thing is an example of not just anthropomorphism but super-extreme-mega-anthropomorphism. It's not just subconscious triggers urging these folks to treat dolphins as if they had human-like minds. No, advocates are saying that they've thought long and hard about this subject in a very conscious way and decided that dolphins are *more* intelligent than humans. These folks would surely deny that anthropomorphism is some nefarious cognitive bias secretly deluding them; rather, it is a well-considered method for justifying their stance on the nature of dolphin minds. They've observed the dolphins' behavior toward pregnant women and concluded that the best explanation for it is that dolphins possess a deep understanding of the human birthing process arising from a kind of superhuman intelligence.

So what advice *did* I give to that good-hearted pregnant couple who asked me to help them with their dolphin-assisted birth plans? Maybe you're imagining that I came barging in with my science stick and smacked the anthropomorphism out of them. But that's not what happened at all. I will tell you what I said, but not before you hear about anthropomorphism's convoluted history as a tool for doing good science. Modern scientists are trained to be wary of anthropomorphism's unconscious influence on our interpretation of animal behavior, but it's not universally considered a bad thing. Not by all scientists, anyway. In fact, scientists are still all over the map when it comes to the potential scientific utility of anthropomorphism. To understand how your average animal behavior scientist living in the year 2024 would tackle the question "What do dolphins understand about the human birthing process and should I ask one to be my doula?," we need to travel back in time 150 years to 1874, to see how old-timey scientists might tackle the problem. The question of when it's OK to anthropomorphize animals has been a topic of debate going back at least that long. And scientists have had wildly different answers to that question throughout history.

The Trouble with Scorpion Suicide

It was the summer of 1874 when a young naturalist named George Romanes first met Charles Darwin in person. The two had begun corresponding after Romanes published an article a year earlier in which he praised and defended Darwin's theory of natural selection.[17] Darwin, a sucker for flattery, invited Romanes to his home in Kent, and the two struck up a lifelong friendship. At that time, the study of animal behavior was the domain of "naturalists" like Darwin and Romanes—early scientists who didn't yet have formal training in an existing academic methodology for the study of animal minds or animal behavior. Instead, the nascent field was catalyzed by Darwin's informal approach to the subject, which he outlined in his book *The Expression of the Emotions in Man and Animals*.[18] It's mostly filled with observations of animals in their natural environment and Darwin's guesses as to what they're thinking using human psychology as the basis for inference. "Young orangs and chimpanzees," wrote Darwin, "protrude their lips to an extraordinary degree...when they are discontented, somewhat angry, or sulky."[19] Since humans feel sulky when they pout, chimpanzees must feel the same, argued Darwin. This approach was later termed anecdotal cognitivism or anecdotal anthropomorphism.

After years of friendship and fruitful discussions about animal behavior with Darwin, Romanes would go on to publish a book in 1879 titled *Animal Intelligence*, which would attempt to dovetail Darwin's anecdotal anthropomorphism into the emerging scientific field known as comparative psychology. Romanes described his landmark book as "a text-book of the facts of Comparative Psychology, to which men of science...may turn whenever they may have occasion to acquaint themselves with the particular level of intelligence to which this or that species of animal attains."[20]

Romanes explained that his method hinged entirely on the idea that drawing an analogy between a human and animal mind via anthropomorphism was the only logical—and thus scientifically valid—way to explain animal behavior. He wrote that "common sense will always and without question conclude that the activities of organisms other than our own, when analogous to those activities of our own which we know to be accompanied by certain mental states, are in them accompanied by analogous mental states."[21] In other words, if an animal is observed acting like a human, it must therefore think like a human. It was an unequivocal defense of anthropomorphism as a tool for science.

For example, he recounts an anecdote published in the scientific journal *Nature* by the famed physician Allen Thomson. While visiting the thermal baths of Lucca, Italy, Thomson was having to deal with the "frequent intrusions" of little black scorpions in the home he was staying in with his family. The best method for removing the scorpions was (according to the locals) to trap them under a drinking glass, wait until nightfall, and then, when it was dark, suddenly produce a candle and hold it near the glass. This would invariably cause the scorpion to freak out and run around with "reckless velocity" before going quiet and, in a moment of existential crisis, stab itself in the head with its own stinger and die.[22] This was termed "scorpionic suicide."[23] Thomson's explanation was that trapped scorpions had the capacity to understand the nature of their plight, lament the inevitability of their impending fate as the flame of the candle approached, and opt to take their own life in order to minimize their future suffering. It's a series of exceedingly complex, human-like thoughts that seemed incredible even to Romanes. But because Dr. Allen Thomson was apparently a "high authority" (that is, an educated white man) Romanes took his word for it and concluded that scorpionic suicide was absolutely a thing.

Other naturalists active at that time, however, thought both that (1) scorpionic suicide was not a thing and (2) Romanes was an idiot for believing it. Among the critics was Conwy Lloyd Morgan, a magnificently bearded psychologist who got into very public science fights with Romanes about scorpion suicide. In a somewhat comical series of rebuttal letters to *Nature*, Morgan explained that the idea of scorpionic suicide was nonsensical and ran counter to the basic tenets of the theory of natural selection. He was so annoyed that he began experimenting on scorpions himself—subjecting them to "sufficiently barbarous" suicide-inducing situations and remarking that none of them went on to stab themself in the head. When people wrote to *Nature* to complain that his experiments were cruel, Morgan wrote back with a passive-aggressive retort: "I am sorry that my experiments on scorpion suicide has given pain to some of your correspondents."[24] Sorry not sorry.

Morgan, like Romanes, wanted to help legitimize the new science of comparative psychology. But he believed the best way to do this was through "carefully conducted experimental observations," not the accumulation of anecdotes.[25] Experiments would produce objective descriptions of animal behavior, and Morgan rejected the idea that you then needed to infer the presence of complex, human-like mental states to produce that behavior. Before believing that animals had minds similar to ours that produce their behavior, Morgan demanded verification of this fact. And since verification of the subjective states of animals was impossible, anecdotal anthropomorphism could never be considered science. "In the psychology of animals no such verification is possible, and verification is that which makes science science" was how Morgan phrased it.[26]

Morgan is perhaps most famous for what would later be known as Morgan's canon. Despite Darwin and Romanes's attempt to take anecdotal anthropomorphism mainstream, it was this edict written by Morgan that would become the bedrock of comparative psychology for the next sixty years: "In no case may we interpret an action as the outcome of a higher psychical faculty, if it can be interpreted as the outcome of the exercise of one that stands lower in the psychological scale."[27]

In other words, whenever you see an animal do something human-like, you should always assume that it is being driven by simple psychological mechanisms (like learned associations) as opposed to complex, human-like cognition (like abstract reasoning). Generations of scientists in the field of comparative psychology lived and died by this rule, afraid that if they assumed the presence of human-like minds in their animal study subjects they would be roasted by fellow scientists waiting to mock the field of comparative psychology as touchy-feely non-science produced by scorpionic-suicide apologists.

To see how Morgan's canon operates and why it sometimes works brilliantly (and thus became the de facto rule for so damn long), consider the mating behavior of turkeys. When a male turkey is courting a potential mate, he will fan out his iridescent tail feathers and perform a ritualistic dance. If the female finds this display convincingly enticing, she will allow the male to mount her. It's easy to draw parallels between the ways turkeys and humans court a mate through dance. Clubs across the globe are filled with human males dancing seductively in the direction of people they find attractive, and science shows that human male dancing ability is correlated with attraction and perceived quality as a mate. One study found that for men, a handful of dance moves were most strongly correlated with attraction, including "variability and amplitude of movements of the neck and trunk, and speed of movements of the right knee."[28] There's a hot tip for all the young men on the dance floor: Fast right-knee movements are what will get you invited in for a nightcap.

If you were to consider turkey dancing through the eyes of Darwin or Romanes, you could conclude that turkeys dance for their mates because, like humans, they are intending to convince females that they are worthy mates. Turkeys, like humans, could very well be using theory of mind to make guesses as to what their female dance-targets find attractive and then dance accordingly. That's what anecdotal anthropomorphism would say.

But Morgan's canon warns that this anthropomorphism-driven assumption of human-like intentions for turkeys is likely blinding us to reality. And this is indeed the case. In a famous study from the 1950s, researchers Martin Schein and Edward Hale from Pennsylvania State University wanted to know exactly what it was about female turkeys that activated the male turkey's lusty dance response. They hypothesized that there was a simple visual stimulus of some kind triggering the male's (presumably mindless) dance behavior. So they took a dead, taxidermied female turkey and placed it in a pen with male turkeys. The males danced for the dead female with the same enthusiasm they'd displayed for a living turkey, and eventually mounted her taxidermized body. Schein and Hale then began methodically removing pieces of the female turkey—wings, feet, legs, and so on—and tested to see if the males still mounted her. By golly, they did. Eventually all that was left was a female turkey head on a stick. This too caused the males to dance and mount the remnants of the taxidermied temptress.[29]

It's hard to imagine a human getting all hot and bothered by a human head on a stick in the middle of the dance floor. This suggests that turkeys aren't evaluating their dance partners using complex, human-like social cognition involving something like theory of mind. They seem to be involuntarily triggered to dance at the mere sight of a female turkey head.

Morgan himself didn't advocate for the idea that all animal behavior was driven by this type of simplistic, turkey-head-on-a-stick response. He didn't outright reject the idea that animals could have complex inner lives, and in fact advocated for the idea that when evaluating experiments like this, the jumping-off point could very well be the question of how a human would react in this situation.[30] But he emphasized that there's no need to jump to the conclusion that complex cognition is the explanation if a simpler explanation is available. That's the thrust of Morgan's canon.

Soon after Morgan's death in 1936, a new wave of researchers was gaining prominence with a novel take on comparative psychology: the outright rejection of the need to study or even consider the possibility of animals having inner mental lives. This was the beginning of the era of behaviorism.

Behaviorism can be understood as the study of animal (including human) behavior without needing to posit the existence of internal mental states. Behaviorists were fundamentally hostile to the anecdote-based approach advocated by Romanes and Darwin since it used the mind of the human observer as the jumping-off point to explain animal behavior. The psychologist Edward Thorndike's 1911 book *Animal Intelligence* was an influential critique of anecdotal anthropomorphism and helped pave the way for the experiment-based approach of behaviorism. Thorndike argued that anecdotes tell you next to nothing; it's experiments that are the key to good science. Thorndike also introduced the highly influential law of effect, which states that "responses that produce a satisfying effect in a particular situation become more likely to occur again in that situation, and responses that produce a discomforting effect become less likely to occur again in that situation."[31] This was his explanation for how animals learned: They simply repeated actions that felt good and avoided actions that felt bad, building up a repertoire of behaviors over time.

You've likely bumped into this approach to the study of animals at some point. It's the explanation behind Pavlov's famous salivating-dog experiments, where a dog can be conditioned to salivate when hearing a bell if you ring a bell whenever you give the dog food. Or B. F. Skinner, who said that "all behavior can be explained as the result of learned associations between a stimulus and a response, reinforced or extinguished through reward and/or punishment." Skinner criticized fellow scientists who entertained the notion that behavior was created by thinking, writing that "the fascination with an imagined inner life has led to a neglect of the observed facts."[32]

The behaviorism craze of the 1950s and 1960s—with its focus on documenting external behavior—became the standard approach used by psychologists to study animal behavior, with university psychology departments filled to the brim with lever-pulling rats in cages being tested on their ability to learn novel and complex behaviors through reward and punishment—a process called operant conditioning. Throughout this period, the idea that animal behavior needed to be explained by complex, human-like cognitive processes like consciousness, inferential reasoning, or theory of mind was anathema. Even Thorndike's law of effect was suspect since it assumed that animals could experience "discomfort." Just because a rat in a cage looked uncomfortable when given an electric shock didn't mean that they were consciously experiencing discomfort in the way a human would. For behaviorists, Morgan's canon was the law, and anecdotal anthropomorphism could be summarily dumped on the trash heap.

But not all animal behavior scientists working at that time were pulled into the orbit of behaviorism. Remember Konrad Lorenz of *Kindchenschema* fame? Lorenz, together with Niko Tinbergen and Karl von Frisch, was awarded the Nobel Prize in Physiology or Medicine in 1973 for research leading to the creation of a new field of study involving animal behavior: ethology. Unlike behaviorism, ethology assumed the presence of instinctive behaviors, immutable predispositions baked into animal brains that evolved to help them navigate their particular ecological niches. These instincts cropped up without needing to be learned and, more importantly, could not be modified by things like operant conditioning, which was a direct challenge to the basic tenets of behaviorism. Ethologists preferred studying animals in natural environments, where these instinctual behaviors had evolved. The ethological approach led to major breakthroughs in the study of animal behavior, like von Frisch's discovery of the famous bee waggle dance, via which honeybees communicate the location of yummy flower patches. Ethologists did not, however, necessarily believe that animals had minds, let alone human-like minds. Like the behaviorists, ethologists believed, as Lorenz stated, that "behavior is to be explained by means of its underlying physiological causal basis, but not by means of invoking psychological drives and subjective motivations of behavior."[33]

But then along came Donald Griffin. Made famous for his discovery of bat echolocation in the 1940s, Griffin weighed in on the anthropomorphism debate when he published a book titled *The Question of Animal Awareness* in 1976. After decades' worth of scientific articles published by ethologists detailing the complex behavior of wild animals, Griffin had simply had enough of the Morgan's canon bullshit, writing that "the flexibility and appropriateness of animal behavior suggest both that complex processes occur within their brains, and that these events may have much in common with our own conscious mental experiences."[34] The book goes on to cite examples of animals doing super-smart things, including the bee waggle dance, but also the ability of great apes to learn and use aspects of human sign language.

Griffin was annoyed that scientists had, for so long, dismissed the idea that animals might have human-like mental experiences as being anthropomorphic. "It is actually no more anthropomorphic, strictly speaking, to postulate mental experiences in another species than to compare its bony structure, nervous system, or antibodies with our own," he argued.[35] For Griffin, anthropomorphism, then, was not some bogeyman that scientists need always avoid, but a scientific tool that could help us better understand animal minds. He defined cognitive ethology as a branch of ethology dedicated to the study of the minds of non-human animals through both research and experimentation. The difference between cognitive ethology and both ethology and behaviorism was that the door was now open for the possibility that animals might have thoughts, beliefs, rationality, and maybe even consciousness.[36]

Griffin would go on to describe those scientists unwilling to entertain the ability of animals to have thoughts and feelings as suffering from mentophobia (an aversion to the idea that animals could have conscious experience).[37] He sparked a powerful pushback against the old-school doctrine of Morgan's canon, arguing that it's just as unscientific to deny the possibility of human-like cognition to animals as it is to assume its presence. The primatologist Frans de Waal coined the term *anthropodenial* to describe scientists' potential "blindness to the human-like characteristics of animals, or the animal-like characteristics of ourselves."[38] The evolutionary biologist Gordon Burghardt introduced a method for studying animal minds in 1985 that he called critical anthropomorphism.[39] This was an attempt to make anthropomorphism not just a "keep an open mind" approach but a true method for studying animal behavior and cognition. In an article unveiling the idea, Burghardt wrote that scientists should use data "from many sources (prior experiments, anecdotes, publications, one's thoughts and feelings, neuroscience, imagining being the animal, naturalistic observations, insight from observing one's maiden aunt, etc.). But however eclectic in origin, the product must be an inference that can be tested or, failing that, can lead to predictions supportable by public data."[40]

Unlike the classic or uncritical kind of anecdotal anthropomorphism offered by Romanes or Darwin, critical anthropomorphism wasn't about taking people's (i.e., old white men's) word for it. It was meant to generate—and test—hypotheses concerning the continuity of the human and animal mind. The primatologist Jane Goodall embodies this modern view. She famously defied scientific convention when studying chimpanzees in the Gombe in the 1960s, giving them names (as opposed to numbers) like David Greybeard, Goliath, Frodo, and Fifi. At the time, her colleagues chastised her for being anthropomorphic. "They objected—quite unpleasantly—to me naming my subjects," explained Goodall, "and for suggesting that they had personalities, minds and feelings. I didn't care."[41] Goodall was an early proponent of critical anthropomorphism, decades before Burghardt had even developed the notion. "Just because you feel that an animal has a humanlike characteristic you cannot assume that is the case," argued Goodall. "Intuition alone is not enough—but it is a wonderful basis for further questioning, testing, and ultimately proving yourself right or wrong."[42]

In her book *Animal Madness: Inside Their Minds*, science writer Laurel Braitman describes the modern, sympathetic take on anthropomorphism (as championed by Goodall and Griffin) as follows: "We can choose...to anthropomorphize well and, by doing so, make more accurate interpretations of animals' behavior and emotional lives."[43] Ecologist and author Carl Safina has argued the same point, stating in a 2015 TED talk that "attributing human thoughts and emotions to other species is the best first guess about what they're doing and how they're feeling, because their brains are basically the same as ours. They have the same structures. The same hormones that create mood and motivation in us are in those brains as well."[44] And so science seems to have come full circle on the problem of anthropomorphism. It was once seen as the perfect tool for studying animal minds. Then it was the worst possible thing scientists could do. Now it seems like the perfect tool again.

But not to everyone.

The Great Divide

Remember the dog scientist Clive Wynne from Chapter 1? Wynne is not a fan of the modern use of anthropomorphism in science.[45] "Anthropomorphism comes very naturally to human beings," wrote Wynne in 2007. "However, anthropomorphism must be resisted. Its drawbacks remain the same as they have always been:

mentalistic folk-psychological accounts of animal psychology have no useful role to play in a modern objective science."[46]

Fellow dog cognition researcher Marc Bekoff disagrees and places himself in the opposing camp. He writes that "scientists who are skeptical about research on animal thinking typically criticize it for being anecdotal and anthropomorphic. Wynne favors reductionistic stimulus–response explanations over ones that appeal to such notions as consciousness, intentions and beliefs. However, he doesn't offer any scientific support for his position. And in fact there is no empirical evidence that the explanations he favors are better for understanding and predicting behavior than those he eschews."[47]

When it comes to anthropomorphism in the study of animal cognition, scientific hot takes are currently clustered at these two extremes. Wynne prefers a "lean" interpretation of the data, where most animal behavior—like dogs' button use—can be explained by simple learned associations, as opposed to complex cognitive skills like theory of mind or consciousness. Bekoff prefers a "rich" interpretation, believing that complex (and thus human-like) cognitive skills can sometimes (but not always) best explain the complex behaviors we are observing in animals. Bekoff has described these differences in viewpoints as "the Great Divide."[48]

The challenge for those scientists hoping to linger in the middle of this divide is figuring out which of the two explanations best appeals to the parsimony principle: the idea that we need to choose the simplest explanation that best fits the evidence. Both camps appeal to parsimony when making their case for or against anthropomorphism. For the rich interpretation, parsimony means assuming that animals with similar brain structures due to a shared evolutionary history (like chimpanzees and humans) would most likely also have shared cognitive capacities. As de Waal puts it: "By far the simplest assumption regarding the social behavior of the chimpanzee...is that if this species' behavior resembles that of ourselves then the underlying psychological and mental processes must be similar too. To propose otherwise requires that we assume the evolution of divergent processes for the production of similar behavior."[49] For Wynne, it's still more parsimonious to assume a bunch of simple cognitive mechanisms clumped together to explain intelligent-seeming animal behavior than to assume the presence of complex, human-like mental states.[50]

As a middle-of-the-divide scientist, I can say that there is no obvious answer as to which of these two appeals to parsimony is most likely to be correct; it depends on the specific situation and the quality of the experiment or observation in question. Science has reached a stage in the study of animal behavior where almost any behavior we witness in animals can be eloquently argued as adhering to either the rich or the lean explanation, with experimentation unable to give us clear answers.

Vengeful Orcas

Take, for example, the strange phenomenon of orcas (killer whales) attacking boats in the Mediterranean. There was a period in the summer of 2023 when multiple news outlets contacted me to ask why the orcas off the coast of the Iberian Peninsula were "attacking" sailboats. Since 2020, orcas have been destroying the keels and rudders of sailboats—chewing on them, ramming them at high speed, and snapping them off. These encounters sometimes last for over an hour as the human sailors huddle terrified on the boat, powerless to stop the mayhem. Even blaring heavy metal music from the "Metal for Orcas" playlist on Spotify created by sailors specifically to deter them only seems to get the orcas even more riled up.[51] Over two hundred incidents have been reported, with major damage to the vessels and at least three boats sunk by the orcas.

Journalists wanted an explanation of what these orcas were thinking when they chewed on and rammed the rudders and keels. The answer that I and most every other scientist interviewed gave was: We don't know. But there were a couple of explanations that came down on either a rich or a lean interpretation of the orcas' possible motivations. The lean explanation was that the orcas had simply learned that keels and rudders are fun objects to play with—things that are satisfying to push around and maybe even break off. And given the evidence that orcas can learn new behaviors from watching each other, the keel-destroying behavior spread through the population as a novel form of play. So the orcas were not attacking boats but just having a grand old time playing with them. Orca experts viewing video evidence of these encounters all seem to agree that the orcas are not acting particularly aggressively toward the boats. It just seems that way to the humans on the boat as they are tossed around like rag dolls. "For them, snapping off a rudder is not really a big deal," I told the CBC. "It'd be like us snapping a Pop-Tart in half."[52]

The second, rich explanation (and the one the media seems to prefer) is that once upon a time, an orca had been injured by a sailboat keel, and now the orcas were banding together to exact revenge on all sailboats they came across. A kind of collective grudge that the resident orca population had acquired made them hate sailboats. "I definitely think orcas are capable of complex emotions like revenge," the director of the Orca Behavior Institute, Monika Wieland Shields, told NPR. "I don't think we can completely rule it out."[53]

It's not outside the realm of possibility for orcas to harbor such thoughts, although this interpretation was not the one preferred by most scientists interviewed on the subject (including Shields). In an open letter written by thirty orca scientists asking people to pump the brakes on the whole revenge thing, the scientists stated: "We urge the media and public to avoid projecting narratives onto these animals. In the absence of further evidence, people should not assume they understand the animals' motivations. Science cannot yet explain why the Iberian orcas are doing this, although we repeat that it is more likely related to play/socialising than aggression."[54]

The reality is that we just don't, and never will, know why these orcas are messing with sailboats. A team of orca experts conducted an exhaustive investigation of these incidents and released a report in 2024 that heavily favored the play hypothesis.[55] "These really shouldn't be called attacks because they're not attacking the boats," explained orca expert John Ford, who was part of the investigation team. "They're playing with the boats."[56] Despite what feels to me like the final word on the phenomenon, both explanations—revenge and play—are at least plausible. And both occupy different ends of the Great Divide.

The one thing I know for sure about the Great Divide is this: Every year, new research comes out with evidence suggesting that animals have more human-like cognition than we once thought. This is a truism in animal cognition science. The rich interpretation is becoming more and more prevalent in the scientific literature, and thus more acceptable to the average animal cognition scientist. "The pressure to avoid anthropomorphism at all costs has lessened," comparative psychologist Joshua Plotnik explained to Katherine J. Wu of *The Atlantic*.[57] Whether or not this means that (critical) anthropomorphism is now accepted as a useful scientific tool, however, remains up for debate. Perhaps the pendulum will swing back toward Morgan's canon once again, although I don't think it will. The thinking, feeling cat is already out of the bag.

This is where we're at when it comes to anthropomorphism's utility for the study of animal minds:

1. It's impossible to ever know objectively what's going on inside animal minds, and using your own mind as a jumping-off point to make guesses is not the worst idea.
2. Some experiments give rather compelling evidence that animals do indeed have human-like cognition some of the time.
3. In the end, it's almost personal preference as to whether you prefer the rich or the lean interpretation of what's causing complex animal behavior, since scientists are still fighting about which interpretation best fits the evidence.
4. Anthropomorphism remains a problem for both scientists and laypeople when dealing with interpreting animal behavior because those subconscious triggers are always lurking in the background, causing us to treat animals as fellow humans even when the evidence is clear that they are not thinking about things the same way we are.

The Ocean Birth Conclusion

So what did I say to that pregnant couple who emailed me about dolphin doulas? I didn't explain that you can't make Soviet superbabies or psychic toddlers or whatever by exposing them to dolphin echolocation, and that the modern scientific evidence I was familiar with made it clear to me that dolphins don't actually give a crap about human pregnancy or newborns. And I didn't mention that male dolphins are known to kill their own calves, so they could just as easily kill a newborn human baby if the mood struck them.

Instead, I simply said that wild dolphins live in the ocean, where there are also sharks. And sharks are attracted to blood, which is something that comes out of people when giving birth. And that both newborn babies and placentas are blood-covered, bite-sized yum-yum snacks in the eyes of a shark. So giving birth in the ocean was essentially chumming the waters to attract giant, carnivorous, baby-eating fish. Only I said all this in a nice, sciencey-sounding way. I honestly believe that giving birth in the open ocean is dangerous and I wanted to help this young couple avoid a potentially deadly situation. They totally got what I was saying and opted not to try for an ocean birth.

I knew the shark thing would work because people fear sharks. And the reason they are so terrified of sharks is … anthropomorphism. Only, it's the flip side of anthropomorphism, the side that makes us *fail* to see human-like characteristics in animals when those characteristics are actually present. The next chapter will explain why we perpetually underestimate the cognitive abilities and misinterpret the behavior of non-cute animals that don't trigger our anthropomorphism.

CHAPTER 3

WALLY THE ALLIGATOR

Why We Misunderstand Ugly Animals

> A person who has good thoughts cannot ever be ugly. You can have a wonky nose and a crooked mouth and a double chin and stick-out teeth, but if you have good thoughts they will shine out of your face like sunbeams and you will always look lovely.
>
> —*Roald Dahl*[1]

Joie Henney's roommate liked to play peekaboo. "If I don't pay him any attention in bed, he'll pop his head out from underneath the covers and give a little hiss or gurgle, and then he'll go back underneath. Then next time he does it, I'll boop him on the nose and then he'll come up beside me and start cuddling."

Joie is talking about Wally, a six-foot, seventy-pound American alligator that lived with him from 2015 until 2024.[2] Wally was rescued together with a batch of other wild "yearlings" from a pond in Florida, but his rescuers noticed something peculiar about Wally. Unlike the other yearlings—all of whom had spent the first eighteen months of their lives living and hunting in the wild—Wally didn't try to bite anyone. "You bump a gator on the nose, I don't care if he's been around humans or whatever, you bump a gator on the nose, you have a great chance of getting bit," explained Joie, who has been rescuing and handling reptiles for thirty-three years. This was the first he had heard of an alligator that wouldn't bite. "He just refuses to bite anything else that is alive and we cannot figure out why."

Because of his bizarre, non-bitey behavior, Joie decided to keep Wally in his home in Pennsylvania instead of giving him to a local reptile rescue center, as he normally would. Every day he tested Wally's biting response by running his hands around Wally's face—and even stuck his hands *inside* Wally's mouth—to see if he could trigger a response. But Wally never once clamped down. Instead, it was quite the opposite: Wally appeared to seek out gentle affection from Joie. "Wally is famous for his hugs and kisses," says Joie.

Wally and Joie lived a rather unconventional life as far as alligator/human relationships go. They were essentially roommates. Since alligators prefer to defecate in water, Joie installed a small pool in his living room for Wally to swim and poop in. Wally rarely had outside-the-pool toilet accidents—even when traveling (yes, I asked). The two of them spent most of their time cuddled up on Joie's couch or in his bed, watching *The Lion King* or *Gladiator* (Wally's favorite films) or playing peekaboo.

When Joie was going through cancer treatment, Wally became his primary source of comfort. "He just constantly stayed in the bed and on the couch with me, cuddling. Even when we were riding in the car, he comes over and puts his head over on the back of my shoulder." When Joie explained his relationship with Wally to his physician, they suggested that Wally might qualify as an emotional support animal (ESA). It's a straightforward process to register an ESA in Pennsylvania, requiring only a letter from a licensed medical professional explaining how the animal benefits the patient during their treatment. Despite being a reptilian carnivore, Wally made the cut, and would wear a little red ESA vest when he went out in public.

Yes, that's right, Wally made plenty of public appearances. When I spoke with Joie in December 2023, he was getting ready for a trip to Philadelphia with Wally: "We were busy almost every day in October and we're booking up for next year already. If everything goes right, looks like we're going to San Diego for a wedding." Wally was going to be the ring bearer—and not for the first time.

Wally made regular appearances at malls, sporting events, and festivals, where people could hold and hug him. Joie's social media accounts are laden with videos of people beaming with joy as they embrace Wally—holding him as tenderly as they would a human toddler. And Wally appeared to be hugging them right back.

Listening to Joie talk about Wally, what he says is indistinguishable from the way pet owners gush over their dogs. Or how Abe Lincoln talked about his Civil War cats. It's unabashedly anthropomorphic. "He means a lot to me," Joie said. "Actually, he means as much to me as my children."[3]

Their bond, as Joie describes it, was a two-way street, involving just as much affection emanating from Wally's side of the relationship. "He knows his name. He knows my voice. Wally senses people's emotions. That's totally unheard of. I've been handling reptiles all of my life, and I'm going to be the first one to tell you they don't care how you feel. They can't care less. But Wally has proven over and over and over again that he has feelings, and he wants to show you affection, and he does it all the time."

When I explained the story of Wally to my friend Briana, she was perplexed. "Aren't alligators just living dinosaurs? Don't they have small prehistoric brains and are driven by basic instincts like food, sex, survival, and sleep?"

That's kind of how I pictured them as well. Not instinct machines per se, as that is the kind of old-fashioned idea about animal cognition that died out during the post-behaviorist, cognitive ethology revolution that we learned about in the previous chapter. And I was vaguely familiar with some research into the intelligence of alligators, crocodiles, and caimans (collectively referred to as crocodilians) suggesting that they've got a fair amount of cognitive flexibility going on. But as for being affectionate, social creatures? I was with Briana. That sounded like a stretch.

Alligator Love

Some reptile experts have accused Joie of misinterpreting Wally's behavior. Chris Gillette and Gabby Nikoll are two professional animal handlers with decades of experience working with crocodilians. In a video discussing the case of Wally, Gabby and Chris are critical of Joie's interpretation of an alligator's capacity for affection. "Because [alligators] are not able to voice how they actually feel, we put our own emotions and our own interpretation of how they feel onto them. And because most people don't understand these animals, [they] could say, 'Oh my God, he loves me so much, and look, he loves my cuddles and he loves my kisses.'"[4] One of the commenters on their video agrees, writing that "Wally's owner has anthropomorphized his alligator, reptiles do not have human emotions."

At first glance, this does seem like a case of anthropomorphism gone wrong—a fundamental mismatch between what an alligator really thinks and feels and how human admirers interpret their behavior to match our hopes and expectations. An affectionate alligator? A playful alligator? An empathic alligator? It sounds nigh on delusional. But when I dug into the science of alligator cognition, Joie's story began sounding more and more plausible. All this anthropomorphism might have turned up some unexpected truth bombs about alligator minds.

I wrote to zoologist and reptile behavior expert Vladimir Dinets to get his take on Wally. I'll be honest: My first thought was that there was something wrong with Wally, a cognitive or developmental delay that caused him to be abnormally gentle and non-bitey. I was expecting Dinets to confirm that Wally was indeed a broken alligator. But I was wrong.

"I don't think you can call such behavior a pathology; it's more like a tail of personality curve," Dinets explained. "Crocodilians have very distinctive personalities. Some are playful and friendly, while most are not."

There have been scattered reports of crocodilians that appear to tolerate—and possibly enjoy—the company of humans. In Costa Rica, a man named Gilberto "Chito" Shedden found an injured adult crocodile that he took home and nursed back to health. Chito released the crocodile, called Pocho, back into the river behind his house, but Pocho didn't leave. He stayed near Chito's home, where the two of them would tussle and play in the river together, eventually becoming a local tourist attraction. At seventeen feet long and weighing over a thousand pounds, Pocho easily could have killed Chito, but the two swam together for over twenty years without any violent incidents.

"A little-known part of Pocho's story is that after his death," Dinets told me, "his human friend spent years working with other crocodiles but never got them nearly as tame and safe to play with as Pocho had been."

Like Wally, Pocho seems to be on the extreme end of the friendliness spectrum as far as crocodilians go. But does this non-aggression really translate into something we can label affection or friendliness? "It is worth noting that wild crocodilians can form long-term relationships among themselves," explained Dinets. "Friendship hasn't been well documented (probably because nobody tried), but it's known that alligators can have preferred sexual partners and would seek them out every mating season. So establishing long-term bonds isn't a completely novel behavior for them. They would also play with each other sometimes, so it's not totally surprising that they can learn to play with humans."

There is actual science to back up these claims of crocodilian sociality, playfulness, and intelligence, some of it published by Dinets himself. Let's start with the playfulness thing. Did Wally really play peekaboo? As unlikely as this sounds, evidence suggests that it's entirely possible.

An oft-used definition of play was provided by Gordon Burghardt (who originated the notion of critical anthropomorphism): "Play is repeated, seemingly non-functional behaviour differing from more adaptive versions structurally, contextually, or developmentally, and initiated when the animal is in a relaxed, unstimulating, or low stress setting."[5] In other words, if an animal is repeatedly engaging in what appears to be an utterly pointless behavior when they've otherwise got some downtime (for example, not hunting, hiding, or mating), then it's probably play. Nobody really understands why play evolved in animals, but it likely helps burn off energy during periods of understimulation and keeps an animal physically and mentally fit. In the case of young animals—who engage in the most play—it might also help generate and develop novel or creative behavior, or otherwise hone skills that are needed in real-life situations.

Like most animals, young crocodilians are more likely to engage in play than adults. They've been observed "repeatedly sliding down slopes into water," giving each other piggyback rides, and surfing waves breaking on shore (not unlike dolphins).[6] Crocodilians enjoy playing with objects, and zookeepers will provide things like logs or balls for them to toss around and chase. A number of crocodilian species appear to have a strong preference for playing with flowers. Pink ones, specifically. As Dinets noted in his research, "Anecdotal observations suggest that crocodilians are generally attracted to small pink objects, and prefer them over similar objects of other colors for biting and manipulating." Crocodilians have been observed gently carrying flower petals on their snouts, or balancing them on their teeth. A recent study found that wild crocodiles in the Smashaan region of India were attracted to the yellow marigolds left by mourners at the river's edge during funeral practices.[7] The crocodiles would spend hours sitting next to (or on top of) the flowers or floating near them in the water, seemingly happy just to be in the presence of the little yellow blossoms.

Wally became famous not for playing with objects (which he also does) but for playing together with Joie, a kind of complex social interaction we typically associate with mammals or smart avian species like crows or parrots. Again, there is evidence to support the idea that crocodilians, despite their reputation as loner cannibals, have the capacity not just for forming individual long-term relationships between mates (as Dinets noted earlier) but also for engaging in communal group interactions. They have been observed hunting together in groups, similar again to the kind of complex hunting techniques observed in dolphins. In one study of American alligators (the same species as Wally), sixty or so alligators were observed coordinating their efforts to corral fish.[8] One group of gators would form a semicircle around a school of fish, driving them toward the shallows, while a second group of active hunters would dart in and grab a meal from the now tightly packed ball of panicking fish. Individual alligators would switch roles (driver versus hunter) over the course of the hunt, seemingly waiting their turn.

As for intelligent behavior, crocodilians have been observed using tools to hunt using a method that feels eerily (and terrifyingly?) humanish.[9] Some species of birds, like egrets or herons, will build their nests near the edges of rivers, collecting sticks from the area. During nesting season, in Florida, there are many species of wading birds that build their nests near the edge of rivers, where alligators live. Sticks are sometimes in short supply, and the alligators have learned to exploit the birds' desperation. They have been observed collecting sticks and balancing them on the top of their snouts, hoping to lure a nest-building bird. They'll then float over to the edge of the river and wait patiently for a hapless bird to try to take one of the sticks strategically positioned near the crocodile's jaws, and then snap them up for an easy meal. Fingers crossed that these alligators don't use this same technique to lure Floridians to their death by balancing iPhones or dollar bills on their snouts.

Cuteification

These observations of crocodilians acting in playful, intelligent, and social ways make Joie's claims that Wally likes to play peekaboo and enjoys cuddling seem plausible. But here's the big question hanging over this story. In previous chapters, we've heard how cute animals trigger our anthropomorphism. But Wally, despite all his supposed cuddliness, is the opposite of cute. In fact, from a *Kindchenschema* perspective, crocodilians have a clear anti-cuteness design. They possess a handful of morphological characteristics that generate an ancient fear response in *Homo sapiens*. For starters, they don't have a round forehead or any soft, chubby facial features. There's no button nose or oversized Disney eyes. They have weird pop-up eyes that protrude out of the top of their head, with no visible whites (sclera)—the hallmark of humanish eyes. Instead, they have an alien-green iris with that reptilian vertical slit, which gives off ominous Eye of Sauron vibes. They have sticky-outy, elongated teeth and a long snout, classic morphological characteristics of scary, people-eating species (wolves, lions, bears, hyenas, piranhas). Their heads are vaguely triangle-shaped, something known to be an ancient fear trigger for our species and found in the deadliest of snakes (viperids).[10] And they have clearly visible, raised scales on their bodies, another visual cue scientifically proven to induce feelings of repulsion in humans.

In one study of humans' fear response to reptiles, crocodilians were singled out as exceptionally scary, far scarier even than deadly venomous snakes. "Morphologically, they fall within the category of other lizards with legs, but compared with them, crocodiles trigger more intense fear in humans," wrote the authors of this study. "Such a result fits well with the fact that crocodiles together with snakes are the only reptilian predators capable of killing a human."[11] Another scientific survey confirmed that crocodilians are near the bottom of the list of potentially cute animals. The only species that ranks worse on the cuteness scale is great white sharks.[12]

So how is Joie not freaked out by Wally? How is Wally bypassing all Joie's fear triggers? I asked cuteness expert Joshua Paul Dale what was going on here. "It's because of cuteification," Dale told me. "If something non-cute is made to look cute or perceived to be cute (often through anthropomorphism), then it can trigger a cuteness response. The leathery faces of turtles are not usually cute. But if a turtle is enjoying a good neck scratch by a zookeeper and appears to be smiling, we'll instantly say: 'How cute!'"

It's easy to see how Wally has been subjected to cuteification. There are online images of Wally flipped upside down on Joie's lap, getting adorable belly scritches, with Wally's mouth slightly open. If you squint just the right way, the ever-so-subtle upturn of his jaw looks like a smile. This is a permanent feature of the alligator mouth—that frozen half-smile (much like a dolphin's smile). It's the same frozen facial expression Wally wears in all those pictures of him cuddling with toddlers at malls across Pennsylvania. In these contexts—intimate cuddle situations where we usually find cute babies or puppy dogs—it's possible to see how crocodilian cuteification involving that reptilian half-smile can override those ancient fear triggers.

Young at Heart

But there's something more to the story. When it comes to Wally, there is a huge difference between the way he *looks* and the way he *acts*. It's his playfulness that leads to anthropomorphizing. Because play behavior is most frequently observed and associated with youth in humans (and other species), it generates a cuteness response. My cat Oscar was extremely playful as a kitten, spending most of his waking hours batting around toy mice, or getting all riled up chasing a piece of string. Now that he is older, he doesn't play as often. But when he does go bounding after a piece of string again, my heart melts ever so slightly. Why? Because he seems so kitten-like, and kittenish behavior is, as we now know, a form of cuteness that triggers an anthropomorphic, caregiver response. So when we see videos of Wally bounding after Joie at the park or flopping onto his back to get belly rubs, it trips that "playful baby" switch.

But playfulness is more than just cute. It's a potential indication that an animal might possess cognitive traits we associate with fellow humans—specifically, those traits associated with agency, like the ability to pretend, to imagine, or to understand the intentions of one's play partners. When human children (or adults) play together, there is often the need to understand the intentions of the other participant. If two people are, for example, chasing a soccer ball around a field, humans are able to understand that the other person "wants" to get the ball and is "trying" to kick it into the net. We imbue our play partners with humanish thoughts and intentions. For humans, then, intention-reading and theory of mind are wrapped up in many forms of play. There are videos online of cows playing soccer with groups of humans, chasing after the ball and seemingly trying to keep it away from the humans. Now, it's possible to go with the lean interpretation of the cow's behavior and suggest that the cow doesn't really grasp the rules of keep-away as they pertain to its understanding of the intentions of the humans. The cow could just enjoy kicking a ball around, and the humans just happen to be there. But to the human observer, it *appears* as though the cow understands the rules. And it's those appearances that generate the anthropomorphic response.

In the case of dogs playing, a rich interpretation of their behavior might suggest that they really do have the ability to understand each other's intentions in a human-like way. Dog play bouts often begin with the play bow posture, with the dog's haunches up in the air, their front paws splayed out, and their head dropped down to the floor. This is a signal that says, "What happens next is play, and not me being authentically aggressive." If a dog should accidentally bite too hard during a play bout, they will again adopt this play bow position, as if to signal, "It was not my intention to hurt you with that bite and I still do want to play."[13] Whether you go with a rich or lean interpretation of animal cognition when it comes to animal play, the result for the human observer is the same: These animals *are acting like* they have human-like minds, with human-like agency and intentions. Our anthropomorphism detectors can't help but sound the alarm.

A crocodile chasing a ball or Wally playing peekaboo generates that same anthropomorphic response to playfulness. Researchers have speculated that the reason we failed to notice until recently the fact that crocodilians are capable of playing is because they, being reptiles, simply don't move as often as mammals. And when they do move, they tend to move more slowly.[14] Their movements were just not human enough to trigger an anthropomorphism response. But, as Gordon Burghardt observed, if you speed up a video recording of the play behavior of Komodo dragons, it looks identical to how dogs play together.[15]

Wally's behavior is more than just playful; it appears to be affectionate, and Joie believes it is. Evidence suggests that at least some crocodilians are (albeit rarely) affectionate with both each other and with their human companions. Affection itself is another behavior that triggers an anthropomorphic response similar to play. When a dog plops its cute little head onto your lap and looks up at you with those pleading eyes, it certainly seems as if they are hoping that you will reciprocate their feelings of affection and maybe want to scratch or stroke them in return. In that sense, affectionate exchanges appear predicated on a two-way understanding of each other's desires, subsequently triggering an anthropomorphic response involving intention reading or theory of mind. Does a crocodile (or a cat or a dog) truly understand anything about their human companion's intentions or desire when it comes to soliciting petting? This again depends on that lean versus rich interpretation of animal cognition. These animals might only consider us humans to be mindless, animated scratching posts that are activated by placing their heads on our laps. In that sense, they have no need to guess at our inner thoughts (i.e., the lean interpretation). But our subconscious anthropomorphism doesn't give a crap what they are really thinking. It *seems to us* that they both understand and are interested in our intentions. And that's how anthropomorphism sneaks in the back door.

Baby Shark

Even sharks can sneak in that back door when it comes to behavior triggering our anthropomorphism. Cristina Zenato is a conservationist and shark diver nicknamed "the Shark Whisperer." She is famous for her ability to remove fishhooks (by hand) from wild sharks. One of her more famous shark rescuees is Foggy Eye, a Caribbean reef shark living off the coast of the Bahamas. After approaching Zenato as she was kneeling on the ocean floor, Foggy Eye allowed Zenato to reach into her mouth and yank out the hook that was buried deep in her throat. That was the moment that Foggy Eye's personality changed, from a curious shark to a hyper-affectionate shark. "As soon as she came back, she allowed me to pet her," explained Zenato.[16] "Since that day she has been the most 'cuddly' of the sharks. She started to come up to me and lay in my lap and has never stopped liking being touched."[17]

Foggy Eye, like many of the sharks Zenato works with, is now fully committed to getting her head scratched whenever Zenato is in the water. She clearly enjoys the feel of Zenato's gentle touch and will rest her head in Zenato's lap for extended periods of time, closing her eyes as Zenato caresses her. It's indistinguishable from how my favorite poodle (Ned) used to rest his head on my lap hoping for (endless) head scritches.

Even the largest and deadliest of sharks enjoys this kind of gentle affection. There is an adult tiger shark in the Bahamas named Emma—fifteen feet long and weighing over a ton—who is world famous for her love of being stroked by human divers like Zenato. "On a daily basis, this shark follows me around like a dog just trying to get her head rubbed," explained Jim Abernethy, who regularly dives with Emma.[18] "These creatures, that most of the world believes to be a mindless, man-eating monster, [are] actually beautiful, sentient creature[s] that thrive on love and affection."

You'd be forgiven for giving this statement the side-eye. A *loving* shark? But there is no denying that sharks like Emma and Foggy Eye do enjoy affection. It is simply impossible to view a video of these sharks with their heads resting on the laps of divers and not feel the anthropomorphism surge in your mind. Like Wally, these animals have a suite of morphological characteristics that should freak us out. Instead, their behavior melts our hearts. Such is the power of human-like or childlike behavior in eliciting anthropomorphism.

Motherf*cking Wasps

We have now seen how humanish behavior (e.g., affection, play, sociality) can override even the strongest yuck response to animals like Wally, Foggy Eye, or Emma. But what about those species that neither look nor behave in ways that at all resemble humans? The ones that don't even make our list of potentially cute animals because they are so obviously grotesque? Things with multiple eyes and limbs, like spiders; or no limbs or visible eyes, like worms; or eight-armed, three-hearted, nine-brained octopuses?

Anthropomorphism instills in us an unwritten rule that the more human-like an animal looks or behaves, the more likely it is that they will have a human-like mind. As we learned in the Introduction, this seems to be the reason anthropomorphism evolved in the first place—as a human-mind-detection device. If it looks like a human and quacks like a human, it's probably a human. But this cognitive bias means that we humans, scientist and non-scientist alike, are less likely to consider the possibility that a creepy, distinctly non-human-looking animal (like spiders, worms, or mollusks) could have a human-like mind.

Consider the story of one much-maligned creepy-crawly: the wasp. With enormous compound eyes that take up most of their heads, huge mandibles, deadly stingers, vein-filled transparent wings, and spindly legs with claws on the end, wasps are full-on nightmare fuel. There's nothing humanish about them. And seemingly nothing humanish about their behavior either. Charles Darwin was so weirded out by wasp behavior that he used them as an argument against the existence of God. Specifically, he was referring to the parasitic wasps of the superfamily Ichneumonidea, which lay their eggs inside the body of a living caterpillar. After hatching, the larvae eat the caterpillar alive from the inside out. "I cannot persuade myself that a beneficent and omnipotent God would have designedly created the Ichneumonidæ with the express intention of their feeding within the living bodies of Caterpillars," lamented Darwin.[19]

But once you get past the disturbing parasitism and morphological weirdness of wasps, there are humanish features to be found—specifically, their (rather adorable) capacity to recognize each other's faces. The golden paper wasp is a bog-standard-looking kind of wasp found in North America. They live in colonies of not more than two hundred individuals. Each individual has distinct markings on their face, which these wasps use to identify each other.[20] If you take a wasp from their nest and paint their face into a different pattern, the colony won't recognize them anymore and will kick them out of the nest, as if a stranger had walked through the front door. This species of wasp appears to have a specialized cognitive system designed to deal with faces—able to process a face as a distinct visual unit (just like humans do), as opposed to learning individual facial features.[21] There's even evidence that wasps (and bees) might be able to differentiate between human faces. So if you feel like a wasp is targeting you specifically because it carries some grudge against you, it's not unthinkable.

Whenever an animal does something that is "like a human," our anthropomorphism—and the media—sits up and takes notice. So when this wasp facial recognition story broke in 2011, news sites went gaga for it: "Like humans, the paper wasp has a special talent for learning faces," proclaimed one headline.[22] "Wasps are as good as humans when it comes to recognising each other's faces," said another.[23] Or, perhaps straight to the heart of the anthropomorphic incredulity issue: "Motherf*cking wasps recognize each other by their faces."[24] We are genuinely amazed when an animal that doesn't look or (normally) act like a human does something even remotely humanish.

Mammalcentrism

The reality is that anthropomorphism has led both the general public and animal cognition scientists to look for human-like cognition in those species that either look the most like us (chimpanzees) or act the most like us (chimpanzee, again). It means we've been slow to search for, find, and acknowledge human-like cognition (or human-like behavior) in non-humanish species like wasps. "How people perceive different species' capacity for sentience is thought to be directly related to how much they differ phylogenetically from humans," argued the authors of one scientific article on why science has been slow to acknowledge the presence of complex cognition in animals like reptiles. "By nature, we humans are drawn towards other mammals, and are better able to empathise with, and accept sentience in mammalian species, than we are other taxa, primarily due to our familiarity with them and the similarities in behaviour and physiology."[25] One study that surveyed fourteen years of scientific research (1994 to 2008) on endangered animals found that there were five hundred times more articles published on mammals than on amphibians.[26] Humans—scientists and non-scientists alike—feel a kinship with other warm-blooded, fuzzy, milk-drinking species that tend to look and behave like us. So they end up being the species that receive our academic attention. This is a bias within science called mammalcentrism.[27]

I asked Vladimir Dinets what it has been like for him as a scientist studying reptiles to deal with mammalcentrism. I wanted to know if it was difficult for him to get the public to accept the evidence that reptiles can have individual (and even affectionate) personalities or engage in play behavior. "I actually found the general public much more open to the idea than many biologists," he explained. "Many scientists have been taught that anthropomorphism is always bad and all animals must be considered machines unless proven otherwise."

Oddly, it might be the case that scientists—those individuals expertly trained to be aware of the pitfalls of anthropomorphism—are actually highly susceptible to anthropomorphism subtly guiding their thoughts when it comes to their choice of study animals. They are victims of the cognitive bias that makes non-human-looking species seem less likely to possess human-like cognition, and thus less interesting or worthy of study. Consequently, work on them is less likely to receive grant funding.

For a great example of this kind of thinking, consider the study of consciousness in animals. It's a perfect example of how anthropomorphism has been at the heart of scientists' (and philosophers') choices of where to look for consciousness in the animal kingdom, and why that might have slowed progress in the field.

The philosopher Kristin Andrews recounts this history in an article in *Aeon* showing that these days we've found "surprising and perhaps disturbing cases of consciousness in unfamiliar places—in animals whose lives are largely hidden from us due to their size, morphology or habitats."[28] But this recent progress happened only because of a concerted effort to break from the historical anthropocentric biases urging us to look in the wrong places. It used to be that language—the quintessential human cognitive trait intimately linked with our capacity for rational thought—was considered an inescapable prerequisite for consciousness. But when the ethologist Donald Griffin argued for science to open its mind to the idea of non-linguistic animals' having some form of conscious experience or self-awareness back in 1976, researchers began considering this possibility, and lo and behold, evidence began to accumulate that animals like chimpanzees or dolphins had something akin to human consciousness. Christof Koch and Francis Crick proposed the idea that "higher mammals" might have the neurobiology necessary for conscious experience in a famous article from 1990 that began the search for the neural correlates of consciousness in animals.[29] "Higher mammals," in this sense, typically means primates, the most human-like of mammals. At the start of the twenty-first century, research into the neurobiology of consciousness moved beyond higher mammals to include all sorts of non-mammalian species, prompting animal cognition scholars (and animal welfare and rights advocates) to pen the 2012 Cambridge Declaration on Consciousness, which proclaimed that "non-human animals, including all mammals and birds, and many other creatures, including octopuses, also possess these neurological substrates," and the 2024 New York Declaration on Animal Consciousness, which states that "the empirical evidence indicates at least a realistic possibility of conscious experience in all vertebrates (including reptiles, amphibians, and fishes) and many invertebrates (including, at minimum, cephalopod mollusks, decapod crustaceans, and insects)."[30] So this would include animals like lobsters, crabs, octopuses, squids, and pretty much every insect you've ever come across.

But Andrews points out in her *Aeon* article that even the radical idea that we should look to an animal's neurobiology as the source of consciousness is itself a result of anthropocentrism guiding our thinking on this topic. After all, the unexamined premise was still that human-like neurobiology was the key to consciousness. These days, there are animal cognition scientists who argue that there is evidence of consciousness in plenty of "lower" animals (like insects) based not so much on their neurobiology as on their complex behavior. "Bees, for example, can count, grasp concepts of sameness and difference, learn complex tasks by observing others, and know their own individual body dimensions, a capacity associated with consciousness in humans," argues behavioral ecologist Lars Chittka.[31] The irony is that scientists who've long been nervous about attributing too much human-like cognition to humanish-looking animals because of anthropomorphism have failed to notice that they've also been attributing too *little* cognition to non-humanish animals for the same reason. Using my own personal history of choosing to study dolphins over chickens when given the option between the two, I think it's safe to say that anthropomorphism has been subtly dissuading scientists from researching the minds of ugly, weird, or non-mammalian species.

And this brings us back to Wally the alligator. Wally helped people push back against the biases that anthropomorphism has instilled in us about not just alligators but all ugly animals. It was his playful, affectionate behavior that made him cute. And this cuteness helps us entertain the idea that he might have thoughts, emotions, and intentions that are more human-like than we might otherwise have expected given his monstrous appearance. Unlike button-pressing dogs or dolphin doulas, where we are at risk of *over*-attributing human-like qualities to a cute animal, we might well have been *under*-attributing human-like qualities to ugly animals like Wally. Or to Emma the tiger shark. Or to those alien-looking wasps.

If, even after reading all this mushy talk about snugly ugly animals, you currently find yourself a little weirded out that there are people out there who hug alligators or cuddle sharks, that's entirely understandable. Personally, I would've loved to cuddle Wally. Or pet a shark. Maybe that sounds horrifying to you. That's because, as we will see in the next chapter, anthropomorphic responses are quite idiosyncratic, with our individual psychologies, life histories, and cultural backgrounds playing a huge role in how our innate anthropomorphism manifests in the way we feel about—and relate to—non-human animals.

CHAPTER 4

STOVE SPIDERS

*How and Why We Don't All
Anthropomorphize the Same*

> When I moved to live in the country, I discovered all these bugs in my backyard. I discovered you can do your own safari. Animals are everywhere. Some are more romantic, like tigers and elephants and chimpanzees, and some are less romantic, like earthworms, but they are just as interesting.
>
> —Isabella Rossellini[1]

In the summer of 2023, a group of my friends was asked to help cook chili for a couple of hundred people as part of a food security initiative in our hometown. We were given access to a rural community center that had a huge commercial stove with enough burners to heat five pots simultaneously. Unfortunately, the stove had not been used in months or maybe even years. It was covered in a layer of greasy dust and insect carcasses. After fiddling with the valves to turn on the propane, we were eventually able to fire up the burners. As the heat from the burners began to rise, it caused a panic among the spider population living in the tangled cobwebs lining the stove hood. These were house spiders of the Pholcidae family, the ones with the long spindly legs that some people call daddy longlegs or cellar spiders. In an attempt to escape the heat, the spiders dropped down onto the heads of the volunteer chefs, trailing silk anchor lines behind them. By my count, over fifty spiders were making the plunge.

Upon seeing this army of arachnids materialize seemingly out of thin air and onto the now panicking volunteers, my friend Donovan shrieked at the top of his lungs and fled the kitchen. Everyone else backed away quickly, unsure what to do. Everyone but my pal Harriet, that is.

I quickly turned off the burners, and Harriet and I got to work rescuing the spiders. One by one we grabbed the anchor thread of each spider and carefully transported them away from the crowd of terrified humans. Pholcidae live indoors, so putting them outside would have been a death sentence. So we relocated them to far-flung corners of the kitchen and abandoned storage areas. Harriet had a big smile on her face knowing that not only were we helping the spiders, but we looked like grade-A badasses, nonchalantly moving spiders from room to room while everyone else shrank in terror.

In terms of what constitutes a "normal" reaction in this situation, I'd say that Harriet and I were the weirdos. Donovan's reaction to the spiders was how most human beings would react. Spiders generate a special kind of fear response, and for good reason. Our species has likely been living alongside (and being bitten and envenomated by) spiders since hominids popped into existence. There is research to suggest that evolution left us with an innate aversion to spiders, with infants as young as five months hard-wired to attend to—and recoil from—spindly-legged stimuli that resemble a spider's form.[2] Thus spiders, like snakes, sharks, and alligators, are animals that possess morphological characteristics that our evolutionary history has taught us to avoid. And this fear is usually helpful, as spider bites are not uncommon and can be unpleasant. Thousands of people are bitten by spiders every day on this planet, although almost all these bites result in nothing more than a skin lesion that resembles a mosquito or blackfly bite, and often go unnoticed.[3] To put things in perspective in terms of how warranted this evolutionarily acquired fear is: Spiders kill an average of 7 people per year in the United States, whereas heart disease kills 611,000.[4] Even in Australia, which is famous for its deadly spiders, it's rare to find spider bite fatalities. One of the world's most venomous—and thus most dangerous—spiders is the Australian funnel web spider (of which there are about forty different species), whose bite can kill a human within fifteen minutes. But this deadly and famously bellicose spider has been responsible for only thirteen deaths in Australia since records have been kept, and zero deaths since 1981, when funnel web spider antivenom was developed.[5] Meanwhile, like in the United States, heart disease is the number-one cause of death in Australia, killing over seventeen thousand people annually.[6] So, statistically, you'd be better served by dampening your arachnophobia and ramping up your fear of bacon.

Anyhow, the noteworthy part of this stove spider story is not that Donovan and the other volunteer chefs were freaked out by spiders. It was that Harriet and I were somehow *not* freaked out. I am fascinated by spiders—especially jumping spiders. So what's going on? What's wrong with us?

Just Built Different

I've met plenty of bug-obsessed weirdos who let anthropomorphism wash over them when it comes to insects or other decidedly inhuman-looking animals. These are people for whom cuteness—whether morphological or behavioral—is not required to trigger an anthropomorphic response. Anthropomorphism, you see, is not a monolithic psychological phenomenon wedged into the minds of all people with the same intensity. It is a product of both our individual personalities and the cultures that shape us.

"Brain imaging studies show that cute things quickly capture our attention, but after that there is a cognitive appraisal process," cuteness expert Joshua Paul Dale explained to me. "This is where individual preference comes into play. We all notice cute things because we're hardwired to do so, but what comes next—how we feel and behave—depends on many factors, including our personal and cultural backgrounds as well as our personalities and preferences."

This delayed-appraisal response is activated whenever we are surprised by the sudden appearance of an animal, whether that's a cute kitty cat or potentially deadly spider. This first phase of appraisal is an instant, unconscious reaction to attention-grabbing stimuli that alert our brain to the presence of cuteness (in the case of kitties) or danger (in the case of spiders). Even Harriet and I initially had this "Look out, danger!" reaction to the stove spiders' sudden appearance. But this is immediately followed by a second appraisal reaction that involves both conscious and unconscious assessments of the situation, culminating in our actual behavior. This assessment is derived from our culture's relationship to the animal in question; our experiences with, education, or knowledge about the animal; our cognitive reasoning style; and the extent to which we normally experience feelings of empathy.[7]

Stove Spiders

You might recognize this type of dual-process cognitive response system, since it's similar to Daniel Kahneman's famous distinction between two modes of thinking when it comes to human decision-making: fast/instinctive/emotional thinking (System 1) and slow/deliberate/logical thinking (System 2).[8] Or the dual-mode camera analogy discussed by psychologist Joshua Greene to describe how humans make moral judgments.[9] According to Greene, we have both an "automatic" setting, allowing us to make quick moral judgments based on emotion, reflexes, and intuition, and a "manual" mode, allowing us to deliberate and reason about our moral judgments based on our understanding of rules and acquired knowledge about the world. Like decision-making and moral judgments, anthropomorphism occurs when two modes of thinking (fast and slow) compete to generate our behavior toward the animals we encounter.

Harriet's and my anthropomorphism-inspired spider-saving behavior is largely due to this second, deliberative appraisal process, derived from our peculiar life histories and upbringings. Unlike Donovan, both of us have had years of casual experience observing, studying, and contemplating the lives of spiders. Both of us knew that Pholcidae are fundamentally harmless to humans. And both of us grew up in families where violence toward bugs was unusual and, in my case, actively prohibited by my mother. My thoughts in that moment involved feelings of empathy for the spiders, whose homes I had just desecrated by turning on the stove. I felt that I owed it to them to move them to a safe location. As we'll see in this chapter, there is a cognitive pipeline leading from anthropomorphism to empathy to feelings of moral obligation.

To see what might've happened to Harriet and me to make our anthropomorphism so spider-friendly, let's go back to where it all started: infancy. As we've seen, humans are predisposed to search for the presence of other, human-like minds. We are all born anthropomorphizers. By the age of one, infants are already more likely to look at living things that move on their own, as opposed to objects being tossed around by the wind or tugged down by gravity.[10] Infants are more likely to, for example, watch a fly buzzing around their head than a leaf falling to the ground. This kind of self-propelled, biological motion suggests the presence of an intention-filled mind generating the insect's behavior, and baby brains are wired to notice the difference. Human brains are also fine-tuned to pick out the presence of faces/eyes in the environment, a trait that is present in babies from day one, with newborns predisposed to orient toward human faces or face-like stimuli compared to similarly complex non-facial visual stimuli.[11] In other words, the biological brain software programmed to search for fellow humans in the environment is already bootstrapping its way forward from the moment a baby emerges from the womb, generating anthropomorphic behavior as it goes. But this mind-detection software doesn't manifest equally in all babies.

Some babies appear more likely to see the world as inhabited by human-like minds than others. One experiment tested a group of four-month-olds to see if they understood the intention behind a very simple act: a hand reaching for an object.[12] In the experiment, a baby would watch a hand appear and push an object. Once they were habituated to (that is, got bored watching) this event, they were then shown the same hand appearing from a different direction and pushing either a new object or the same object. If the baby understood that the hand had a new "goal" (that is, it "wanted" to push the new object), the babies would look longer at that event as opposed to the hand appearing and pushing the same object as before. Fifteen of the twenty infants tested looked longer at the event involving the hand's new goal, suggesting that they were surprised/interested by a change in the hand's intentions. Four years later, these same twenty kids were given a verbal test of their ability to understand the desires, beliefs, and intentions of other people—a theory-of-mind test. The fifteen kids who did better on the theory-of-mind test were the same ones who, when they were infants, had looked the longest when the hand reached for the new object. In other words, some of the kids showed an innate predisposition for framing the world as a place filled with other, intention-filled minds, and this predisposition was something they carried with them as they grew up.

Humans, it seems, exist on a spectrum when it comes to their built-in propensity to theorize about the existence of other minds. Based on the study I just mentioned, you'd think there'd be a simple way to understand the relationship between a child's capacity for theory of mind and the extent to which they might, for example, arrive at the idea that spiders have humanish minds by way of anthropomorphism. Perhaps people born extra-sensitive to mind detection are the same ones who freely attribute minds to non-human animals. But it's not that straightforward. A number of studies have found that people with autism are far *more* likely than people without autism to assume that animals or objects have human-like minds.[13] This is counterintuitive since the hallmark of autism is difficulty in understanding both one's own emotions and the mental states (the emotions, intentions, desires, beliefs) of others—that is, a deficit in theory of mind.[14]

Researchers studying this phenomenon suspect that "autistic people may have strong tendencies to attribute mental states as often, or even more often, than non-autistic people, which leads to indiscriminate attribution of mental states to people and objects (i.e., anthropomorphism) alike."[15] In other words, because folks with autism have difficulty correctly guessing the mental states of other humans, they are actually more likely to assume the presence of mental states that are not there in an attempt to compensate for the difficulty. And this leads to willy-nilly mental state attribution, and thus increased anthropomorphizing.

Overactive Anthropomorphizers

There are other personality and psychological traits that are predictors of our likelihood to anthropomorphize. If you had an imaginary friend as a kid or if you suffer from chronic loneliness, you are far more likely to anthropomorphize.[16] For example, if I were to ask you if your favorite shoes get lonely when you don't wear them, most people would say no. But if you are one of the growing number of people experiencing chronic loneliness, you are far more likely to tell me that your shoes do in fact get lonely.[17] Or maybe that your cereal is sad. Or perhaps the spiders living over your stove are scared of the heat. Interestingly, you can generate this same kind of anthropomorphic response by inducing temporary feelings of loneliness in people. In one experiment, participants watched a clip from the movie *Cast Away* (where Tom Hanks's character is isolated on a desert island for four years) and then answered questions about their pets' internal mental states. After watching the clip, the participants were more likely to ascribe to their pets anthropomorphic traits involving social connections, such as being thoughtful, considerate, and sympathetic, than those in the control group who had watched clips from the baseball movie *Major League*.[18] In other words, watching a heavily bearded Tom Hanks weep as his volleyball friend is swept out to sea will subconsciously nudge you into thinking that your cat loves you and wants you to be happy.

Stove Spiders

I don't think Harriet's and my "overactive" anthropomorphism is due to chronic loneliness, temporary loneliness, or autism. It is, however, likely that we've been above-average anthropomorphizers since birth. Combine that with our unique family situations, where showing respect and kindness toward all animals (including creepy-crawly ones) was the norm, and you have the recipe for an adult human who sees spiders as deserving of moral consideration. Both of us grew up in households with multiple pets of various species—a situation known to amplify future anthropomorphic behavior. It turns out that having grown up with a cat or dog will have a far more profound influence on your anthropomorphic behavior regardless of what you learn about animal psychology in school, as will having grown up surrounded by wildlife in the backwoods of Maine instead of in the concrete jungle of Manhattan.[19]

By the time we reach adulthood, both nature and nurture will have coalesced to give us a unique, baseline capacity for anthropomorphism. And this baseline is something that is both scientifically measurable, and able to reliably predict our future anthropomorphic behavior. Researchers Adam Waytz, John Cacioppo, and Nicholas Epley developed a questionnaire (the Individual Differences in Anthropomorphism Questionnaire, or IDAQ) to test how prone people are to anthropomorphism. It included questions like "To what extent do cows have intentions?" and "To what extent does a car have free will?" People tended to answer these questions in remarkably similar ways regardless of whether the subject of the question was an animal or an object.

The higher people scored on the test, the more likely it was that they would anthropomorphize in new situations. The test result, in other words, could be reliably used to predict someone's future behavior, suggesting that it's correctly detecting the presence of a general anthropomorphism capacity. People who score high on the IDAQ are more likely to attribute emotions to non-human animals, find it problematic to harm computers or other inanimate objects, think it is wrong to harm the environment or nature in general, and trust technology like AI or polygraph tests to provide helpful information when making important decisions.

I took the IDAQ myself, and the results were as expected: For all questions involving animals, I gave them a 10 out of 10 for having things like free will, consciousness, emotions, et cetera, and I gave all objects (like televisions, tables, and the ocean) a 0 out of 10 for these things. Despite my scientifically informed belief that non-living things lack any form of cognition, my average IDAQ score was likely higher than most thanks to my intuitive (and also scientifically informed) belief that animals are riddled with complex cognition. The IDAQ results help explain much of my and Harriet's behavior, and clarify why we reacted the way we did that day we tried to make chili and the spiders started dropping onto our heads. It also explains why we both are rather vocal environmentalists: It is one of the expected knock-on effects of being an above-average anthropomorphizer.

Bellamie, Mon Ami

At this point, it's becoming clear that attributing minds to non-human animals via anthropomorphism goes hand in hand with feelings of empathy. There are lots of ways to define empathy in psychology, but for the purposes of this book, let's assume that empathy works like this. First, one must understand that another individual (human or animal) has a mind of its own capable of consciously experiencing emotions. Second, when we observe an individual showing behavior that we guess is a result of their experiencing an emotion, we either simulate that emotion in our minds (on an intellectual level) or actively experience a similar emotion. This then nudges us to feel compassion for the other individual.

To see how mind attribution, anthropomorphism, and empathy can get all jumbled up, I want to share with you a story my friend Jess Lalonde, a jazz singer, told about her relationship with a butterfly. Here's her story, in her own words:

> In the summer of 2020, my relationship with my fiancé fell apart, which also meant saying goodbye to the dog we shared (the dog that I loved and cared for like he was my boy-child flesh and blood). I thought I was doing OK and dealing with my heartache in a healthy way, but it turns out I was shoving things down, living in denial, and pain, anger, and grief hit me like a tidal wave around Christmas 2020.
>
> By summer of 2021 I was out of the dark valley of heartache and sort of standing on the plains of tender forgiveness but also feeling extra fragile, gentle…delicate. I was tending to my pollinator garden at my parents' house when I found a butterfly hopping under the orange sprigs of a butterfly bush with part of his wing hanging by a thread. I kept him safe and made a little terrarium habitat for him. I named him Bellamie [beautiful friend]. Eventually the wing came off.
>
> I fed him mushed banana and honey water from the palm of my hand, enjoying the slurping sensation from his little proboscis. We had our family cottage week planned, so I brought him with me in his terrarium. My bedroom at the cottage was small and had a baby crib in it, so I put

him there at night and continued taking him for supervised outings in the garden.

One time, I accidentally left his cage door open and he wandered out and fell through the cracks of the deck, so I climbed down through the spiders and dirt like Orpheus descending to the underworld and brought him back to the light. Ultimately, Bellamie fell off his perch and drowned in his water dish while I was on a date.

Later I was reflecting on this whole period with my sister-in-law and I asked her if she'd been worried about me. She'd witnessed all of it and spent a lot of time smiling politely and nodding at me. I think I was expecting her to say, "No, it was really sweet," but she guffawed and said, "Yes! I thought you'd lost your entire mind and I couldn't believe your parents were indulging your madness! I didn't know what the f*ck was going on—you put a butterfly in a baby crib!"

The truth is, I felt a deep connection with Bellamie—he couldn't fly, though he kept trying.... He was so, so vulnerable and I couldn't leave him. I felt his delicate little body vibrating in my hand and I didn't see the difference between my beloved dog or family cat or him—his life force was the same. I loved him, and it doesn't take a shrink to see in caring for him I was caring for my slightly tattered self.

This is about the clearest articulation of the process of animal empathy I've encountered. Jess looked at Bellamie's situation and drew concrete parallels to her own mental and emotional state. She imagined that Bellamie was experiencing feelings of loss and sadness, pain and distress that were not just analogous (similar) but homologous (identical in form and function) to her own feelings. And this compelled her to act to alleviate his suffering—to not only anthropomorphize Bellamie but also act on her powerful feelings of empathy.

Jess's relationship to Bellamie is strange as far as the science of inter-species empathy goes. Insects rarely generate this kind of empathic response in humans. But Jess is clearly empathizing with Bellamie to a (in her own eyes) concerning degree—eloquently making the connection between her own feelings of emotional fragility and the damaged, fragile state of a one-winged butterfly. Like me and Harriet, Jess appears to be an overzealous empathizer and anthropomorphizer—likely generated by a similar predisposition that she's had since birth and then a lifetime spent anthropomorphizing. She occupies the extreme end of the animal empathy spectrum.

The Other End of the Empathy Spectrum

There are people who are entirely unable to empathize not just with animals but with other humans. I am talking about psychopaths. Psychopathy is considered a personality disorder that primarily affects one's ability to care about or empathize with other people. About 1 percent of the population fits the clinical definition of psychopathy. People with psychopathy are more likely to be involved in violent crime, with up to 25 percent of incarcerated people in the United States likely to be psychopaths. Psychopaths are often interested only in furthering their own interests, and view other humans as a means to an end. In that sense, they are able to understand on an intellectual level that other people (and maybe animals) have minds, but this understanding does not manifest the ability to imagine or simulate what it would feel like to be another person. This lack of empathy hinders the ability to feel any moral obligation to help others, and makes it instead far easier to harm them without remorse.

What causes psychopathy is not well understood, but psychopaths are known to have a terrible time reading and understanding the facial expressions of other humans. Something in their neurobiology makes it difficult for them to, for example, understand what fear or pain looks like on the face of a fellow human. As you can imagine, this leaves them less susceptible—if not fully immune—to *Kindchenschema* faces. Without cute kitten and doggy faces eliciting anthropomorphism and empathy, psychopaths simply don't care all that much about the well-being of animals. In fact, being "physically cruel to animals" is one of the criteria in the *Diagnostic and Statistical Manual of Mental Disorders (DSM-5)* for a diagnosis of conduct disorder in youths, a common precursor to a psychopathy diagnosis.[20] In one example discussed by Daniel Wegner and Kurt Gray, a teenager diagnosed with conduct disorder was asked to identify the emotion of fear in the facial expression of another person, and explained, "I don't know what that expression is called, but I know it's what people look like right before I stab them."[21]

And it's not just psychopaths. Research shows that other personality traits like narcissism (a clinical diagnosis defined as "a pervasive pattern of grandiosity, need for admiration, and lack of empathy, with interpersonal entitlement, exploitativeness, arrogance, and envy") and Machiavellianism (someone who engages in "interpersonal strategies that advocate self-interest, deception and manipulation") are also associated with a higher likelihood of animal cruelty.[22] Sociopathy, or antisocial personality disorder, is a close cousin of these other disorders, and also involves difficulty with empathy, as well as deceitfulness, recklessness, and impulsivity; it too occasionally results in animal cruelty. The writer, psychologist, and diagnosed sociopath Patric Gagne describes in her book *Sociopath: A Memoir* an incident where she found great pleasure in strangling a cat nearly to death.[23] Thankfully, she stopped—not because she was suddenly overcome with empathy, but because she realized that others might be annoyed that she murdered a cat. She has spent a lifetime learning to control these impulses and is now a successful spouse, mother, and cat owner, having lived not by the guiding hand of empathy but by a set of explicit rules she's made for herself to stop her from harming other people and animals.

Aside from psychopathy and sociopathy, the single biggest personality trait associated with animal cruelty is a propensity for everyday sadism, which describes someone who derives pleasure, excitement, and maybe even sexual arousal from acts of cruelty or suffering.[24]

Incidentally, it's not only deviant psychological traits that are associated with a higher likelihood of cruelty toward animals. There are basic demographic factors that predict one's propensity to harm animals. Young people engage more frequently in acts of animal cruelty than adults. And boys are often more likely to harm animals than girls.[25]

Some Are More Equal than Others

But a blasé attitude toward animal suffering is not always directly associated with pathological personality traits like sadism, psychopathy, or being born a boy. Nor is it always a result of a more general inability to empathize. How we treat animals, and the extent to which we either anthropomorphize them or feel empathy toward them, is enormously impacted by culture. Specifically, food culture.

We are all aware of the fact that different cultures consider some animals fair game to consume but others taboo. In general, we eat pigs but not cats here in North America, unless you are following Jewish or Islamic law, at which point both pigs and cats are off-limits. People in Vietnam, however, eat both pigs and cats. Dog meat is eaten in South Korea (where it is currently the fourth-most-consumed type of meat), although this practice is slated to be banned as of 2027.[26] The ban results from a cultural shift in the perception of dogs in Korea, where younger generations increasingly view them as pets and not food.

For a human to eat the meat of an animal, that animal will need to be killed. The context for the killing of an animal, and the species being killed, will determine whether it's an acceptable thing to do or a sign that someone is displaying deviant psychological traits that could get them diagnosed with psychopathy or sadism. Currently, if I were to kill and eat my cat here in Canada, I might well be headed for a psych evaluation. But not if I were to kill and eat one of my chickens. In South Korea, killing and eating a dog is not currently associated with deviancy or a psychological defect, but try it after 2027 and you'll find yourself in trouble with the law. Our cultural ideas about which animals it's OK to kill and eat, and thus which ones it's acceptable to anthropomorphize, can shift over time.

I had firsthand experience with the diversity of cultural views on the moral status of animals—particularly cats and dogs—in the summer of 2023, when I helped produce a play starring a dozen people who had recently immigrated to the small town in Nova Scotia where I live. The play focused on people's experiences as immigrants and some of the cultural differences we'd encountered since moving to Canada. My friend Samir from Syria wrote and performed a stand-up comedy routine for the show in which he joked that here in Canada (unlike back home in Syria) husbands seemed to be less respected than dogs when it comes to who receives top billing in the family unit. When I told my friend Kulbir from Punjab, a fellow actor in the play, that I was writing a book about anthropomorphism, he had a lot to say on the subject. He explained that back in India, dogs do not command the same respect or receive the same level of affection as they do in Canada. This idea of pampering your pets with fancy dog treats and dressing them in little raincoats was absurd, and a stark contrast to the way street dogs in India are either ignored or reviled. For Samir and Kulbir—as with many people across the planet—these are just dogs, after all. And treating them like fellow members of the family, like little fur babies, was simply bizarre.

Clearly, humans have the capacity to either actively suppress or activate the triggers that generate our feelings of anthropomorphism toward some species depending on our culture's view of them as either food or pets. In order to take an animal's life and throw its body into a crockpot, we must extinguish any empathic response that activates when, for example, we see a cute kitten or wide-eyed Chihuahua. Research into the psychology of slaughterhouse workers shows that they must develop a "coping mechanism that enables them to emotionally disconnect and distance themselves from slaughtering."[27] In the book *Slaughterhouse* by Gail Eisnitz, one slaughterhouse worker described the need to actively dissociate from their feelings of empathy for the animals being killed: "It gets to a point where you're at a daydream stage. Where you can think about everything else and still do your job. You become emotionally dead."[28]

The *Jaws* Effect

But even for the species we don't usually eat, our culture shapes our anthropomorphic response to them. And these cultural attitudes are always in flux. Consider the terrible fallout of the Academy Award–winning 1975 film *Jaws*. After *Jaws* became a global cinema sensation, the public developed an unprecedented fear of sharks that had negative consequences for shark populations. Sharks were transformed—almost overnight—from a relatively unknown, rarely seen ocean dweller that landlocked people rarely thought about to a world-famous ocean monster with an insatiable taste for human flesh. The public was suddenly scared of swimming in the ocean, and the solution seemed to be to kill as many sharks as possible to make the waters safer. "In *Jaws*, the only solution is to kill the shark," writes social scientist Chris Pepin-Neff, who coined the term "the *Jaws* effect" to describe the impact the film had on our cultural understanding of sharks. "It is explained that a rogue shark will continue to hunt for prey (humans) in an area unless the food supply stops or it is killed."[29]

This is, of course, total nonsense. There are no rogue sharks that develop a taste for human flesh—this is a Hollywood fabrication. Not only are shark attacks incredibly rare, but sharks can be adorable little cuddle bunnies, as we saw in Chapter 3. But it was too late—the idea that bloodthirsty sharks needed to be killed to prevent attacks on humans had entered the zeitgeist. Pepin-Neff documented how the *Jaws* effect was still influencing government policy in the early twenty-first century in Western Australia, where the government advocated for the preemptive killing of sharks spotted in areas frequented by human swimmers in order to prevent "imminent" shark attacks. When shark experts pointed out that the sharks living in the area posed absolutely no threat at all, policymakers seemed unable to listen to the science. They simply couldn't shake the images of killer sharks.

The author of the book that the *Jaws* film was based on, Peter Benchley, regretted the impact his book and the film had on people's perception of sharks. "I couldn't write *Jaws* today," he said. "The extensive new knowledge of sharks would make it impossible for me to create, in good conscience, a villain of the magnitude and malignity of the original."[30] Steven Spielberg, the director of *Jaws*, expressed similar sentiments: "To this day, I regret the decimation of the shark population because of the book and the film. I really, truly regret that."[31]

While the story of *Jaws* makes it clear how culture can influence how we feel about certain types of animals, sometimes it's difficult to explain exactly why we've all collectively decided that we like or dislike some species. Take raccoons, for example. I have always been surprised that raccoons have received such little academic attention when it comes to the study of animal intelligence. There are endless studies on the cognitive capacities of chimpanzees, dolphins, parrots, and dogs—but hardly any on raccoons. Scientists just don't seem that interested in working with them, despite the fact that they have, in my opinion, all of the cuteness triggers you'd find in a puppy.

To ask why this was the case, I wrote to one of the few raccoon cognition scientists out there: Lauren Stanton, a cognitive ecologist with the Animal Behavior and Cognition Lab at the University of Wyoming. "There is a lot of anecdotal evidence of their cognition (problem-solving, learning, etc.) but raccoons are generally an overlooked study species," she explained to me over email. Perhaps this is because humans consider raccoons to be annoying, nuisance animals that are mostly seen rooting through our trash and compost bins. "They are really good at taking advantage of opportunities they find (one of the primary reasons we want to study their cognition), but unfortunately this means that raccoons and humans will sometimes come into conflict," explained Stanton. "If a raccoon has denned in someone's chimney or figured out how to get into someone's garbage can or chicken coop, they are not doing this intentionally because they are evil 'geniuses,' they are doing this to survive."

Our collective cultural aversion to these evil little trash pandas has likely resulted in scientists ignoring raccoons as a potential study subject. Stanton's group is one of the few out there looking at the underappreciated intelligence of raccoons. In one study, researchers baited puzzle boxes with prunes soaked in the oil of canned sardines (a raccoon delicacy) to test their problem-solving skills.[32] The box contained a series of latches, locks, and sliding panels behind which the prune could be found. Some of the raccoons managed to solve the puzzle by rapidly cycling through a list of possible solutions until one worked. It was trial-and-error on steroids. "Raccoons are indeed rapid (associative) learners, and they demonstrate flexibility in problem-solving," Stanton explained. But our negative cultural biases concerning raccoons—just like our cultural feelings about sharks—have led scientists to avoid studying these smart little buggers until only recently.

Think like a Deer

All these examples show that culture has a huge impact not just on whether we anthropomorphize but on how that anthropomorphizing happens and what that means in terms of the right or wrong way to treat an animal. The extent to which an individual feels the tug of anthropomorphism when it comes to how they relate to non-human animals (due to their personal psychology, upbringing, and culture) has a direct influence on the moral position they adopt when it comes to the treatment of animals. "If anthropomorphism involves attributing humanlike mental states to nonhuman agents," write Adam Waytz, John Cacioppo, and Nicholas Epley, "then it should also predict the extent to which people consider and respect a nonhuman agent's interests and wellbeing."[33]

It's not always the case that the more one anthropomorphizes, the higher the likelihood that one feels that harming animals is wrong or unethical. I was chatting with the anthrozoologist Hal Herzog (author of *Some We Love, Some We Hate, Some We Eat: Why It's So Hard to Think Straight About Animals*) about this subject. He pointed out that for people who hunt and eat wild animals, there is an inherent paradox in the way they anthropomorphize. "You are more likely to be a good hunter if you can think like a deer," he explained. "So there's a paradox in that the more you can think like an animal—the more you can identify with it—the more morally problematic it should be to kill it."

On paper, adopting the perspective of the deer you're stalking—really getting inside its head and trying to imagine what it's like to be a deer—should lead to greater concern for its well-being. That is, after all, the definition of empathy. Slaughterhouse workers, in contrast, try very hard not to think about the minds of or empathize with the animals they kill. But for almost all hunters, a sophisticated ability to think like a deer doesn't lead to being paralyzed by a moral quandary when staring down the end of a rifle at an eight-point buck. It actually leads to a better kill count. This is a paradox that has likely been with us since the moment our species evolved to think about the minds of the animals we were hunting. Clearly, we must have developed a psychological mechanism that allows us both to anthropomorphize and empathize with the animals we hunt but also to still justify killing and eating them. If we didn't have this ability, our hunter-gatherer ancestors would have switched over to a gathering-only lifestyle as soon as anthropomorphism started to blossom in our minds.

But framing a discussion about the paradoxical relationship between hunting and anthropomorphizing as a matter of psychology highlights a problem with the very nature of this discussion. Many non-Western cultures have a fundamentally different relationship to animals, and therefore show differences in how anthropomorphism manifests and bleeds into a discussion of empathy and moral obligation. The worldview of, for example, the Indigenous Mi'kmaq people of North America does not map easily onto the kind of anthropomorphism-to-empathy-to-moral-obligation pipeline as I've been outlining it. For many Indigenous cultures that hunt animals, there simply is no psychological paradox to overcome.

I currently live in the ancestral home of the Mi'kmaq—called Mi'kma'ki—which today incorporates parts of Quebec and Atlantic Canada. The Mi'kmaq have been hunting moose, caribou, and beaver in this area for millennia and consider these (and all animals) to be brothers or relations. This idea is, in a way, anthropomorphic in that humans and animals are viewed as equivalent when it comes to assumed similarities in how their minds work. "Humans and animals both experience our lives in the first-person," writes Indigenous scholar Margaret Robinson in her review of animal personhood in Mi'kmaq culture, "overcoming fears, having adventures, falling in love, raising families, vanquishing enemies, and having a relationship with Kisu'lk, the Creator."[34] Calling this situation "anthropomorphism" seems almost incorrect. It's something more than that. Within the Mi'kmaq worldview, a human cannot humanize an animal insofar as humans and animals are essentially all "people" to begin with. "Many stories written and oral stories in Mi'kma'ki do often present animals as self-aware rational people," Robinson explained to me over email.

For Western scholars, it might seem odd that Mi'kmaq hunters don't notice the inherent paradox—that, for example, they could consider a moose to be a self-aware, rational person but still hunt, kill, and eat that same moose. After all, in the Western world, rationality and self-awareness are cognitive attributes we often find in ethical and legal arguments for extending rights and legal personhood to animals. "How does a culture that recognizes other animals as persons and relations reconcile this worldview with a diet that is so heavily animal derived?" asks Robinson in her review. "The answer is in the Mi'kmaq belief that animals willingly sacrifice themselves to become food."[35]

Within Mi'kmaq culture, animals will gladly sacrifice their bodies to help their human brethren survive. And the hunters will, in return, treat the animals' remains, as well as the environment they call home, with gratitude and respect. Ultimately, the Mi'kmaq worldview as it pertains to the nature of animal souls is what negates any potential ethical problems for a hunter. "The death of an animal's physical body may be separated from the death of the animal person, whose spirit is believed to go to Waso'q, the Land of the Souls, an afterlife in which human and animal souls coexist in harmony, with all their needs fully met," observes Robinson in her review.[36] Within this ethical system, there is no harm being done to an animal that is willingly sacrificing itself to a hunter and whose soul lives on after death. This example is a handy reality check to remind those of us working within Western philosophical and scientific traditions that not only does culture impact how we anthropomorphize, but the concept of "anthropomorphism" itself might not make any sense within some cultural worldviews where humans and animals are by definition cognitively and morally equivalent.

The Ghost in the Machine

Consider Japan's unique relationship to cuteness and how this likely relates to that society's ancient animistic belief systems. "In Japan, cuteness has had a more prominent place in art and literature than it did in other places for at least a thousand years," Joshua Paul Dale told me. This is perhaps why the modern form of cuteness or *kawaii* culture originated in Japan. "The animist/polytheist nature of Japan's spiritual tradition persists in the way people approach animals and objects," explained Dale. "I'm not saying that everyone still believes that non-living objects have a soul or consciousness, but people tend to *behave* as if they do due to this long tradition. This makes it easier for people in Japan to make connections with cute objects."

The ease with which the Japanese anthropomorphize both cute animals and cute objects as ingrained in their culture is scientifically detectable in even very young Japanese children. In one study looking at the extent to which children (around age five) from Japan, Israel, and the United States attributed humanlike mental properties to other humans, plants, animals, and inanimate objects, the Japanese kids were the most likely to anthropomorphize inanimate objects.[37] This is precisely what the researchers expected given Japan's thousand-year-long tradition of viewing inanimate objects (cute or otherwise) as having souls or consciousness. By the time children reach the age of five, Japanese culture is already clearly influencing their propensity to anthropomorphize. Perhaps buried deep in the Japanese cultural psyche is the possibility that not just animals but all things have a life force or consciousness and that this obliges us to treat them accordingly.

It's here that we begin moving toward the next manifestation of anthropomorphism—away from animals and living things and toward inanimate objects. So far, we've seen that anthropomorphism is rooted in the need to detect fellow human-like minds in our environment. And while animals might not have minds identical to our own, their minds are likely similar when it comes to experiencing things like pleasure and pain. So it's not so strange that anthropomorphism makes us feel morally obliged to be kind to them. It's not ridiculous, in other words, that people like Harriet, Jess, and I go out of our way to help insects in distress. And if our unusually high baseline anthropomorphism brings a little more kindness into the world, that can only be a good thing, no? This is another argument for reframing anthropomorphism as a potential good in this world—it's a behavior that creates more happiness not just for the people engaging in it but also for the living creatures that benefit from it.

But the same cannot be true when it comes to anthropomorphizing inanimate objects. Unless you appeal to a spiritual belief system that like the ancient animistic system in Japan, attributes consciousness to inanimate objects, then any application of anthropomorphism to non-living objects is wrongheaded, a misfire of the mind-detection system. And yet, as the next section of this book will show, we anthropomorphize inanimate objects just as easily as we do non-human animals. I hope to show not only that it is not wrongheaded but also that it is a completely normal, totally fun, and maybe even biologically beneficial thing to do.

CHAPTER 5

COMPANION CUBE

Why We Get the Feels for Inanimate Objects

Mark Watney started talking to his potatoes. He wasn't sure when it had begun, but he found himself giving them pep talks, apologizing for the occasional accidental scorching, and even giving them names.
—*The Martian*[1]

One morning back in the late 1990s, I was walking through the backstreets of Drogheda, Ireland, on my way to catch the train to Dublin. The alleyways were strewn with empty milk cartons—a distinctly Irish form of litter that I have never been able to explain. Among the detritus I spotted a tiny stuffed monkey, a casualty from the discarded McDonald's Happy Meal that lay nearby. I hurried to catch my train but spent the rest of the day thinking about this monkey. Its beady little plastic eyes had been staring up at me as I walked by, and I felt sad that I had left it among the street garbage. By the time I made it back to Drogheda that evening, I was overcome with a feeling that I needed to rescue the little monkey. I found him exactly where I had last seen him—nobody had bothered to pick him up, which made my heart bleed for him all the more. I brought him home to my wife and explained what had happened, which did not surprise her at all. I am known to have a soft spot for stuffed animals—I've literally kept every stuffed animal I've acquired since childhood. We ran the monkey through the washer, and he came out looking good as new. He was given the moniker Laundry Monkey and had the honor of sitting on a shelf above our washing machine—the patron saint of dirty things waiting to be made clean again.

Here's the thing: I am not crazy. I know damn well that this monkey is just a bunch of fabric and plastic and does not have any capacity to feel "sad" at being abandoned on a street in Drogheda. And I know that the empathy swelling in my chest when I first saw the unwanted monkey was generated by psychological mechanisms in my brain that create this sometimes irrational anthropomorphic response. I *know* this. I know this on a professional level, for crying out loud. And yet... I still treated that little hunk of fabric and plastic as tenderly as I would a human baby. So what is going on here?

Laundry Monkey was triggering an empathic response in me that made me act as if he (the hunk of monkey-looking fabric) had thoughts and feelings, an internal mental life of some sort that I needed to respond to. I didn't want him to be sad, so I picked him up and took him home. Clearly, these triggers were happening below the level of consciousness—I hadn't reasoned my way into this bonkers position. Even when I consciously ruminated on the irrationality of these feelings, I still seemed unable to stop myself from feeling them. I had anthropomorphized Laundry Monkey—seemingly against my will.

The anthropologist Stewart Guthrie once described anthropomorphism as an "involuntary perceptual strategy" that is a "product of natural selection, not of reason," which perfectly captures the weirdness happening in my mind when I brought Laundry Monkey home with me.[2] Similarly, Nicholas Epley described the unconscious triggers for anthropomorphism as a "sixth sense" that operates almost invisibly.[3] We don't need to consciously believe that the thing we're anthropomorphizing actually has an inner mental life identical to a human for our minds to nudge us to treat it like a human. The psychologist Gabriella Airenti argued precisely this point, writing that "anthropomorphism is not grounded in specific belief systems but rather in a specific modality of interaction."[4] We are subconsciously compelled by anthropomorphism to treat cute-faced stuffed monkeys as if they have feelings and mental experiences, even in cases where we might intellectually believe that they lack these experiences.

It's vital to understand that anthropomorphism is an invisible, involuntary process that partially operates below the level of rational thought, as these anthropomorphism experts are contending. This explains why it crops up just as often for objects as it does for animals. As we will see in this chapter and the next, it's those three primary triggers of anthropomorphism—eyes, movement, and language—that are responsible for the bulk of this invisible triggering.

The Eyes Are the Mirror of the Soulless

There's an evolutionary explanation as to why Laundry Monkey's sad little eyes could peer directly into my anthropomorphic soul and activate that anthropomorphism-to-empathy-to-moral-obligation pipeline, just as I had experienced with those stove spiders. As we know, eyes are a powerful anthropomorphism trigger. Not just cute baby eyes of the *Kindchenschema* variety, but *all* eyes.

Humans are hard-wired to notice eyes, and there are cognitive systems at work that oblige us to search for faces and eyes in our environment. Why? Aside from revealing the location of a predator or prey species (which is handy info), noticing a pair of eyes out there in the wild could indicate the presence of a fellow human with whom we might form a social connection. Once that social connection is established, human eyes are designed to convey vast quantities of information about that person's inner mental life.

Internal emotional states are displayed through subtle changes in the position of the flesh and muscles around our eyes, and our perceptual system evolved to detect and interpret these communicative signals. Eyebrows can be contorted into shapes that the human mind interprets as anger, fear, disgust, confusion, and so on. Even the direction of one's gaze is loaded with information. Looking someone directly in the eyes is a cue that you are hoping to initiate a social interaction. But hold that gaze a bit longer and it can be seen as a threat or a possible sign of an impending attack—or, depending on the social context, a sign of lustful intentions. And the direction in which someone is looking reveals where their attention is focused, which helps us guess what they are planning to do next. Because my brain is hypersensitive to the presence of eyes in my environment, even Laundry Monkey's beady little plastic eyes were enough to trigger my anthropomorphism.

The hard-wired nature of the human propensity to look for eyes and faces in our environment and use that information to read intentions and internal mental states is what leads to face pareidolia. Pareidolia is an illusion caused by looking at ambiguous stimuli and our minds forcing meaning onto them. Thanks to face pareidolia, humans are predisposed to having our minds trick us into seeing a face whenever we come across stimuli that bear even a vague resemblance to two eyes and a mouth. The book

How Are You Peeling? Foods with Moods is just a bunch of pictures of fruits and vegetables with googly eyes or maybe some fake teeth glued onto them.[5] But so unstoppable is face pareidolia that we perceive these veggies as humanish beings that are laughing, crying, or otherwise expressing emotions that they 100 percent do not have.

Moved by Movement

Eyes are just one of the main triggers for anthropomorphism that seem to work just as well for inanimate objects as living creatures. Movement is another. We are hard-wired to consider anything that moves of its own accord (self-propulsion) as having goals and intentions stemming from a mind similar to ours; this is a predisposition that manifests in infancy.[6] The more humanish the movements, the more triggered we are to believe they are being controlled by an actual human mind.[7] This is especially true if that movement is random or unpredictable. Humans and other animals have an inbuilt capacity for folk physics—that is, a naturally occurring understanding of how objects behave in the world when encountering things like gravity or wind. If a leaf falls slowly to the ground while being blown around by the wind, this does not make us see intentional movement in the leaf. It's just doing normal leaf stuff.

But if that leaf should stop falling, slowly turn toward us, and then rapidly accelerate in a straight line before flying up into the sky, then we *would* be triggered. This kind of behavior, which involves a leaf defying the laws of physics, is generally only explainable by intentional, thoughtful control, which suggests the presence of an intelligent mind. Seeing a leaf behave like this would make me jump to the conclusion that the leaf was a tiny drone being controlled by a human hidden in the bushes. Or maybe it was a miniature spacecraft being piloted by a tiny alien with human-like intelligence. If you read reports of people who claim to have seen alien spacecraft, it's almost always the movement of the UFO (that is, the appearance of self-propulsion and defying physics) that they cite as evidence of alien visitation. In one study, researchers put wheels on a garbage can that could be driven by a hidden operator with a remote control to see how people would react to seemingly autonomous movement.[8] People tended to ascribe intentions and desires to the garbage can, describing it as "hungry" as they waved it over or whistled for it like a dog so they could throw their trash into it.

The Heider-Simmel illusion is the classic example of how humans see humanish intentions in certain types of movement. In a famous experiment from 1944, the psychologists Fritz Heider and Marianne Simmel showed people a brief animation with three geometric shapes (a large triangle, a small triangle, and a circle) and a box with an opening.[9] No eyes anywhere, just shapes. The shapes moved around the screen, bumping into each other, and going in and out of the box. After viewing the video, people were asked to describe what they saw. Instead of saying things like "The large triangle first was in the box and then moved out, and then the circle moved into the box," they framed the video as a narrative. They'd say things like "The large triangle is a bully and was acting aggressively toward the small triangle and circle. They had to run away and hide from the mean triangle. He was probably jealous." There was something about the movements of the shapes that seemed, to a human, as if these shapes were nervous, scared, belligerent, angry, and so forth. Thanks to those subconscious anthropomorphism triggers, people couldn't help but view the shapes as having intentions and emotions—not because they had humanish facial features or morphology, but simply because their movements were happening in a humanish way, with sudden, unpredictable changes in direction and variations in speed.

Movement *speed*, it turns out, is another anthropomorphism trigger all on its own.[10] Had the shapes in the Heider-Simmel illusion been moving at a far greater speed, the illusion would have been shattered. Animals or things that move faster or slower than a typical human are less likely to trigger our anthropomorphism. This timescale bias, as it's called, is perhaps why we are so weirded out by the distinctly inhuman slowness with which sloths move, which causes us to subconsciously view them as lazy and unintelligent and not particularly humanish in the cognition department. The word *sloth* literally means "laziness," and this word is the name European explorers gave to the animal when they first stumbled across it in the jungles of South and Central America.

You might think that at least one of these two major anthropomorphism triggers (eyes/faces, movement) would have to be present for us to anthropomorphize an object. But the crazy thing about the human mind is that we sometimes detect intentions even in cases where there are no physical anthropomorphism triggers whatsoever. Humans bend over backward to frame the world as rife with intentionality.

Please Take Care of It

To see how easy (and kind of weird) it is for us to see humanish minds or intentions in inanimate objects even in the complete absence of any anthropomorphism triggers, I need to tell you about my favorite video game: *Portal*. In the game, an AI named GLaDOS (Genetic Lifeform and Disk Operating System) has forced you (the player) to solve a bunch of complicated puzzles involving teleporting around a series of test chambers at the fictional Aperture Science testing facility. Sometimes you need to place an object on a specific spot on the floor to trigger a door to open or some other kind of effect. For one level, the game designers created a puzzle where the player was supposed to carry a giant metal box from the start of the level all the way to the end, and use it along the way to solve some of these puzzles. But when they first unveiled this level to their game testers, they found that people kept forgetting to take the box with them.[11]

While researching ways to manipulate the psychology of the player to remind them to take the box with them, one of the game designers, Erik Wolpaw, found a government document on interrogation techniques that suggested that "isolation leads subjects to begin to attach to inanimate objects."[12] This discovery inspired game designer Kim Swift to find a way for the player to form an attachment to the metal box. Her solution was simple. She painted a heart on the side of the box and renamed it the Weighted Companion Cube. When the player first finds Companion Cube, it's accompanied by an instructional graphic (like the ones you see on IKEA furniture assembly instructions) depicting someone affectionately cradling the cube in their arms. GLaDOS then tells the player, "This Weighted Companion Cube will accompany you through the test chamber. Please take care of it."

"After that, no one ever forgot the box," Swift said in an interview.

Later in the game, the designers had to find a way for the player to learn to use an incineration device that would be needed when fighting the final boss. Their solution was to have the player practice using the device by incinerating Companion Cube. After the player arrives at the end of the testing level, GLaDOS says,

Companion Cube

"The Weighted Companion Cube certainly brought you good luck. However, it cannot accompany you for the rest of the test, and unfortunately must be euthanized." A pit in the ground then opens up and the player is meant to throw their Companion Cube friend into the fire. "Although the euthanizing process is remarkably painful," explains GLaDOS, "eight out of ten Aperture Science engineers believe that the Companion Cube is most likely incapable of feeling much pain."

The idea of incinerating the nominally sentient cube was too much for some players. When they first tested the level, some people refused to do it. "A couple of people jumped into the incinerator themselves rather than kill the cube," Swift noted.[13] Thankfully, Companion Cube makes an appearance at the end of the game, much to the relief of those of us who struggled with the idea of killing our cube friend. Companion Cube would go on to become the most loved character in *Portal*, attaining cult status in the gaming world, and spawning all sorts of merchandise. You can buy Companion Cube throw pillows, Companion Cube underwear, and, of course, a sexy Companion Cube Halloween costume.

Based on what we've learned about how anthropomorphism is triggered (or not) when we look at living creatures or inanimate objects, consider how bizarre this situation is for a second. Companion Cube contains exactly zero anthropomorphism triggers. There's nothing humanish about it. No puppy dog eyes, expressive eyebrows, or adorable oversized feet. It doesn't make any cute sounds. It doesn't even move, let alone move in a humanish way like the shapes from the Heider-Simmel illusion. It is just a stationary metal box with a heart on it and someone telling you that it is your friend. And yet, people playing *Portal* form legit emotional bonds with Companion Cube. Or should I say the *idea* of Companion Cube; it is a two-dimensional, onscreen version of a metal box, after all. The game designers even mock the players for their attachment to Companion Cube, with GLaDOS saying at one point that "hallucinations" and "perceiving inanimate objects as alive" are symptoms generated by the stress of playing the game. The game designers are being obvious with their ploy to get the player emotionally invested in the fate of Companion Cube. As the scholar Megan Arkenberg wrote, "*Portal*'s 'marketing' of the Cube proves so successful, in part, because its attempts to make us love it are transparently obvious, and the game's self-reflexive irony ensures that the player never feels 'taken for a sucker.'"[14] The player is in on the joke, as it were.

But how is it that Companion Cube—an object with precisely none of the traditional triggers for anthropomorphism and whose creators are openly joking about the silliness of forming an attachment to it—can nonetheless elicit a real anthropomorphism response? The answer is: The player wants to anthropomorphize the cube. It's fun. There is pleasure to be had in pretending that non-human things have humanish qualities. We allow ourselves to feel anthropomorphism toward Companion Cube because it pleases us. Mind-to-mind interaction is what made our species so successful, so it's not strange that natural selection designed our brains to derive pleasure from the process. We enjoy it so much that we allow ourselves to do it even if we know that those minds are obviously absent, as in the case of Companion Cube. This revelation is a game changer, as it helps explain much of our behavior when it comes to anthropomorphizing not only inanimate objects but also animals and abstract concepts. And it helps explain why there are such huge individual and cultural differences in the way we anthropomorphize.

The Anthropo-Dial

To understand how this works, I need to introduce you to a new concept. I am calling it the Anthropo-Dial. Imagine the Anthropo-Dial as a little dial in your brain that controls the intensity with which you anthropomorphize something. Remember, I am using the word *anthropomorphize* to mean "treat something as if it were a fellow human." The Anthropo-Dial has settings ranging from not-human (we don't treat the thing like a human in any way whatsoever) up to fully human (we treat the thing as if it were another human with a mind identical to our own), with a spectrum of humanish settings in between that result in varying degrees of human-like treatment.

The default setting of your Anthropo-Dial—which is unique to each person—is heavily influenced by all the things we learned about in the first part of this book. As we learned in the previous chapter, some people are born with a greater propensity to see other minds or intentions in the world around us. And some cultures are more likely to anthropomorphize than others. The family one grows up in and the exposure we have to animals, robots, or other humans can also change the default setting. The beliefs you hold about the likelihood that things like animals could have humanish minds—stuff you learned about in school or through discussions with friends—will also reposition the dial. People whose Anthropo-Dial is at a much higher setting than the average person's are more likely to get swept up in anthropomorphizing a butterfly, Laundry Monkey, or Companion Cube.

Then there are all these subconscious triggers for anthropomorphism that we've learned about—all these doggy eyebrows or Heider-Simmel kinds of movement that we know can make us more likely to engage in anthropomorphism. Each time our minds encounter a trigger, it turns the Anthropo-Dial up a few notches, making it more likely we will anthropomorphize the thing doing the triggering. So for people whose dial defaults to a high setting, you can imagine that if they spot some eyes on a banana or a stuffed monkey or whatever, it might just spin their dial to an absurdly high setting, leading them to treat that banana or monkey nearly as kindly as they would an actual human baby.

Crucially, each of us can consciously decide to move our Anthropo-Dial. We can choose to anthropomorphize Companion Cube because it's part of the game we're playing and it would be more fun if we did, and thus we crank up the dial a bit. Turning the dial overrides the default settings, making it easier to engage in the behavior of anthropomorphizing and, in the case of Companion Cube, experience the fun of pretending that a metal cube has feelings. The fact that we know the cube can't possibly have feelings doesn't matter; we can consciously decide to twist that dial to generate anthropomorphic behavior toward Companion Cube, which results in that pleasurable tinge of quasi sadness when we grapple with throwing it into a firepit. It's an illusion that we actively allow our minds to generate.

We can fiddle with the dial in all sorts of situations, although it might not always be easy to turn depending on the intensity of the triggers or the location of our default setting. For those people working in slaughterhouses, we saw that they could twist the dial way down in order to kill animals that otherwise might trigger a fair amount of anthropomorphism. Doing so might allow them to slaughter animals, but if that violent behavior clashes with their default anthropomorphism setting influenced by anthropomorphism triggers and their underlying beliefs about animal minds, they might have to jump through a few psychological hoops to explain or justify their actions to themselves. For some people, like me, the Anthropo-Dial is cranked way up, making it nigh on impossible for me to suppress the pull of anthropomorphism and even imagine working in a slaughterhouse. Turning the dial too far from one's default setting can make anthropomorphizing things (or not) a real challenge.

The Anthropo-Dial thus helps explain both (1) why we can treat things as if they are humanish even though we don't truly believe that they are anything like a human (like Companion Cube), and (2) why we can treat things as if they are not humanish even though we might believe that they are fairly humanish (like slaughterhouse workers can do). We can override our default anthropomorphism settings or any anthropomorphism triggers depending on the situation, and crank that dial as needed. Surely plenty of slaughterhouse workers have pets at home, and they can crank their dial right back up when interacting with their favorite cat. Anthropomorphism is a behavior that we can consciously dial up or down depending on our psychological needs or wants in the moment. This helps explain the seeming cognitive dissonance in the next type of anthropomorphism for inanimate objects that I want to talk about: pretend play.

Fake It Till You Make It

A perfect example of how the Anthropo-Dial functions during pretend play is hobbyhorsing. There is a long and diverse global history of people dressing up as horseback riders, perching behind a fake horse head (or horse skull), and pretending that their legs are the horse's legs (examples are the Kuda Kepang of Java and the Polish Lajkonik). The modern sport of hobbyhorsing makes use of the toy hobbyhorses that many of us from North America would remember from our youth: those wooden sticks with a plush horse's head stuck on the end of it.

The origins of hobbyhorsing as a sport began in Finland in 2012. Young girls in Finland had invented a way of dancing and prancing with their toy hobbyhorses that closely resembled the way Olympic dressage horses could move. They would hold their upper bodies erect as if they were riding a real horse, but move their lower limbs in a manner that looked elegantly horse-like. In online forums and videos that gained traction in the 2010s, these girls would show off their skills, and they would get together in real life to practice and coach each other. Soon they began competing in hobbyhorse competitions that involved dressage, gate jumping, and forest trail rides. These competitions had real judges, real prizes, and real stakes. As videos and documentaries about this new sport spread across the internet, hobbyhorsing went global. There are now numerous hobbyhorsing competitions around the globe, with the largest and oldest held in Finland: the Finnish Hobbyhorse Championships, first held in 2013. This weeklong event now has over eighteen hundred competitors.

Hobbyhorsing is practiced almost exclusively by tween and teenage girls. For many, the absence of boys and adults from the sport is part of the appeal. Many girls will make their own hobbyhorses, carefully crafting the horse heads and imbuing them with personality and unique physical characteristics. When it comes to the anthropomorphic relationship these enthusiasts have with their stuffed horse heads, the line between pretending and believing can be a bit fuzzy. At competitions, real equine veterinarians are present to help the girls check their horses for disease and make sure their vaccinations are up to date. In a video I saw online of a competition, one of the high-jump competitors hugged her horse after landing a jump—in much the same way you'd praise a living horse for executing a difficult jump.

For the girls who take competitive hobbyhorsing seriously, they don't consider what they are doing as pretending at all. "If someone says we are playing, it strips away everything we made," says Alisa Aarniomäki, one of the originators of hobbyhorsing and star of the documentary *Hobbyhorse Revolution*.[15] Sometimes hobbyhorsing enthusiasts talk about their horses in ways that make this fuzzy line between pretending and believing all but disappear. "He is a very gentle horse, he learns fast, and he really loves to jump," said Maisa Wallius about her hobbyhorse Tarzan.

Clearly, some hobbyhorse enthusiasts have turned their Anthropo-Dial up to a super-high setting. But the question is: Do they really, truly believe that their toy horses have humanish thoughts and feelings, or is this, like Companion Cube, just something they are pretending to do because it's fun? In all likelihood, if pressed on whether they really think their hobbyhorse has feelings, every child would admit that no, not really. They might resent you for asking because this makes the game less fun, and, as Aarniomäki notes, it "strips away the reality." But psychologists studying pretend play in children have found that children know the difference between play and reality. "Even when young children anthropomorphize the objects with which they play, they are not confused about their status," writes anthropomorphism scholar Gabriella Airenti. "It has been shown that at least by age 3, children distinguish reality from pretense."[16]

For pretend play, children can turn their Anthropo-Dial way up in order to experience the pleasure of interacting with their hobbyhorse or Star Wars action figure or Barbie doll as if the toys had humanish feelings. But they never seem to tick over into actually believing that these things are real humans. This is thanks to something that I call the Humanity-Limiter, a psychological buffer created by your mind that ensures you reap the benefits/pleasures of treating a non-human thing similarly to a fellow human but never risk deluding yourself into believing that the thing *is* human. Remember, humans have an inbuilt capacity and biological need to chunk the world up into the categories "human" and "not-human," and the Humanity-Limiter is there to facilitate that process during the act of anthropomorphizing. Think of it as performing a function similar to the scald guard in your shower—a device that if you start turning the dial toward the hot water, generates a little resistance or even stops the dial from turning so you don't accidentally turn the dial too far and scald yourself. The Humanity-Limiter is there to stop people like me from turning my Anthropo-Dial too far and actually believing that Laundry Monkey has the same mind and moral value as my wife. The Humanity-Limiter will crop up in future chapters as we deal with confusingly humanish things, so file this concept away for easy reference.

As we know from previous chapters, some individuals are more prone to anthropomorphize than others, and there are parallels to this in one's capacity for pretend play. Children who have imaginary friends are more likely to anthropomorphize in general.[17] And kids (and adults) who readily engage in role-playing are likely to be rampant anthropomorphizers of both animals and objects.[18] From a personal perspective, this seems to check out. I, as you now know, am a rampant anthropomorphizer, and also a Dungeons and Dragons player, where I pretend to be all sorts of creatures, from wizards to bugbears to dragons and even talking tables. I also teach improv, which is essentially just pretend play for adults. If you didn't think I was a supernerd before, I imagine this revelation has clinched it for you.

Anthropomorphic Fictional Characters

Our ability to turn that Anthropo-Dial to humanize non-human things for reasons of play or pleasure goes a long way in explaining why our literature and film is chock-full of anthropomorphic characters. Surely I don't need to explain how utterly universal these kinds of characters are, but just to be sure, here is a short list of famous books and films featuring either animals or objects that are humanish in their behavior and/or morphology: *The Wind in the Willows, Watership Down, Thomas the Tank Engine, Animal Farm, Beauty and the Beast, Winnie the Pooh, Stuart Little, The Brave Little Toaster, Babe, Guardians of the Galaxy, Planet of the Apes,* and *Toy Story*.

Why is it so common for us to depict in our art animals and objects as having humanish qualities? The answer may be that humanish characters are better positioned to convey a narrative because they remove the danger of fully connecting emotionally or morally with a human character. "Anthropomorphism, animal characters as people, can add a degree of emotional distance for the reader/writer/speaker when the story message is very powerful, personal, and painful," wrote education scholars Carolyn Burke and Joby Copenhaver. "Having animals do the acting and mistake-making allows the face-saving emotional distance often needed to be able to join the conversation."[19]

To dig deeper into this subject, I sat down with English professor Kailin Wright, who teaches a children's literature course at St. Francis Xavier University. "I ask students to think about why anthropomorphic animals are so popular in children's literature on the very first day of class," she explained. "I think that what is really there with anthropomorphism is being able to accept that animals are like us, but ultimately animals are not like us." The humanish but not fully human status of a fictional character (like Charlotte, the talking spider from *Charlotte's Web*) bumps up against the Humanity-Limiter, allowing a child to immerse themself in the story without the danger of getting too attached to a real, human character. "It doesn't have the same risk of exclusion as seeing a real child," Kailin noted. Regardless of a child's ethnic, racial, or cultural background, they can empathize with—or maybe even see themself reflected in—Charlotte's struggles. This "like us but not like us" idea is one of the reasons Charlotte's eventual death is easier for a child to process than if Charlotte had been a human character. Because she is in the category of non-human, we don't have the same inbuilt ethical obligation toward Charlotte as we would a real human.

There is evidence to show that children do sense a difference in ethical weight when it comes to a fully human character versus a humanish one. When researchers tested to see if stories featuring human characters or anthropomorphic characters were better at teaching children moral lessons, it was the human stories that worked best. "Contrary to the common belief, realistic stories, not anthropomorphic ones, are better for promoting young children's prosocial behavior," the researchers concluded.[20]

This difference in moral weight is what facilitates the emotional distance between us and our beloved fictional anthropomorphic characters. Strangely, though, this distance can also increase the strength of the actual emotions we feel. It's as if the story gives us the freedom to let strong emotions wash over us because we know that these emotions are being applied to a not-really-human character in a fictional situation where nobody can be hurt for real.

For example, the saddest moment in cinema history for me was when, in the animated film *Toy Story 2*, Jessie the cowgirl was abandoned in a cardboard box by the side of the road while Sarah McLachlan's tear-jerking ballad "When She Loved Me" played in the background. Had this same scene played out as live action, I am not sure my reaction would have been as strong. A real human child version of Jessie being abandoned by her family might have been more likely to generate an unpleasant trauma response in me and send my thoughts spiraling down a path of moral intellectualization to justify or explain what I was seeing onscreen, or perhaps spur me to dissect the behind-the-scenes filmmaking choices of the director or cinematographer in an attempt to shield myself from the unpleasantness of it all. Thus, I might have actively suppressed my feelings of sadness. But thanks to the Humanity-Limiter preventing me from ascribing true human status to Jessie, I could instead set aside all of that complex thinking and simply wallow in the pure sadness of innocent Jessie's anthropomorphized abandonment. I explained to Kailin how this scene affected me, and she recounted a similar reaction to the moment in the film *Cast Away* when Tom Hanks's volleyball, Wilson, floated out to sea. "I really cared about Wilson," she admitted.

Does the Dog Die?

I suspect these strong reactions to the simple, honest, blank-slate innocence of Jessie and Wilson are related to a similar phenomenon involving the portrayal of animals in literature and film. I was just watching an episode of the TV show *Fallout* where one of the main characters (The Ghoul) stabs and almost kills a dog. We had just seen The Ghoul kill a dozen humans moments before, with graphic depictions of their deaths. But the stabbing of the dog happened offscreen. Why? The editors were likely wary of showing violence against a dog because some audience members would find it unpleasant—more unpleasant than the violence against humans they had no qualms about putting onscreen. It's why that infamous dog-killing scene in the film *John Wick* is the scene everyone singles out as being horrific, and not the seventy-seven humans he murders throughout the film. And even that traumatizing dog death happened offscreen.

My friend Ashley recalls walking out of the theater during a screening of the film *I Am Legend* when Will Smith's character had to kill his dog after it got infected with the zombie virus. "The moment he had to kill it, I was like, 'I'm fucking done with this movie. I'm leaving. Good night. Goodbye.' I left and I got in my car and I cried for twenty minutes," she recalls. At this point in the book, you might be wondering why I have so many friends who cry in their cars about dogs. Due to so many moviegoers' aversion to onscreen animal violence, a website was created, called Does the Dog Die?, where you can search for a film or TV show to find out if it contains any scenes that animal-violence-sensitive people might find triggering.

I asked the political sociologist Loredana Loy, who has studied the portrayal of animals in film and television, about this aversion to onscreen animal violence. Loredana pointed out that context is key, and there is plenty of onscreen animal violence that is used for comedic effect, like the gruesome fate of Aunt Bethany's cat, who gets electrocuted after chewing on the Christmas lights in *National Lampoon's Christmas Vacation*. In cases like these, the audience is often *not* encouraged to anthropomorphize the animals, so the scene comes across as funny as opposed to horrifying. But it doesn't take much to frame the animal character as one that deserves our empathy and thus encourages anthropomorphizing—similar to how Companion Cube was made part of the protagonist's story in *Portal*. "If the animal is part of the story in some way," she explained to me over email, "as opposed to just being used in some dramatic way in a fleeting scene, the likelihood that they will be given moral consideration—regardless of the species—goes up."

In the examples I mentioned so far, each of the animals or inanimate objects was indeed an important character in the story. The animals in films that we are the most emotionally involved with often display, as Loredana puts it, the "perception of helplessness." This helplessness is clearly related to the idea of these characters being anthropomorphized as moral patients—like how we respond to real-life babies, or how Abe Lincoln responded to those kitty cats. When combined with the animal's role as a character in the story, helplessness ensures the audience's anthropomorphic response to them. But thanks to the Humanity-Limiter, we do not ascribe the full suite of moral and cognitive traits to either onscreen animals or anthropomorphic characters and can instead enjoy the pleasure of sobbing at the fate of Jessie the cowgirl or John Wick's dog.

The Benefits of Anthropomorphizing Objects

We have seen that it is fun and pleasurable to anthropomorphize objects or anthropomorphic fictional characters for reasons of play or storytelling. But our capacity to anthropomorphize non-living things might have evolved for biologically beneficial reasons that go beyond pleasure. To illustrate this point, I need to introduce you to a couple of my object-loving friends.

My jazz musician friend Jake has an emotionally powerful and clearly anthropomorphism-inspired relationship with his Gibson ES-335 Dot 1958 Reissue guitar: a not-uncommon phenomenon for a professional musician. The way Jake talks about his guitar, it's as if his Anthropo-Dial is cranked about as far as it can go before his Humanity-Limiter breaks and he falls into delusion about the human status of his Gibson.

Jake has had this relationship with his guitars ever since his very first guitar. When his first guitar was stolen, he went through a mourning process. "It was like losing a family member," Jake told me. "I had lost both my grandparents on my father's side and one of my grandparents on my mother's side. So I recognized that emotion." Jake was given a replacement guitar, and that produced a similarly strong anthropomorphic response. "It kind of felt like I was cheating on my real guitar with some other guitar." The new guitar—the Gibson ES-335, which he named Chantelle—is the one he has formed the strongest bond with. "One day Chantelle fell from the stand and then hit the floor and the headstock of the guitar snapped off. I don't think I've ever felt so much shock or horror in my life."

My friend Alasdair has experienced something similar during his time as a sailor. "Boats are storied in a way which few objects are," he wrote to me when I asked him how he feels about the boats he worked on. "Ask anyone who has sailed on an old traditional sailing boat for a length of time and get ready for a few hours of stories. You form a relationship with a boat. You are moved by it."

The affection sailors develop for their boats leads them to experience a sense of obligation toward them—what Alasdair called a moral weight. "I once remained as mate on the tall ship *The Eye of the Wind* for months longer than I initially agreed to (or should have), out of a sense of duty to the ship. I remember one engineer, Terry Pooley, an older gent, returned to the ship many years after he first sailed on her out of a sense of duty to the 'old girl.' The job was not well paid, but felt he owed some service to the ship itself." To note here is that Alasdair used the pronoun "it" to refer to the ship, while Terry Pooler used "she." There is a long tradition of gendering boats as feminine in English (going back to at least the fourteenth century), with the apocryphal idea (according to *Legion Magazine*) being that ships are seen as "motherly, womb-like, life-sustaining vessels, protecting and nurturing their crews through good times and bad."[21] As we saw in the history of the study of animal behavior, ascribing a gendered pronoun to a non-human (whether that's a chimpanzee or a boat) can generate feelings of anthropomorphism, which is why it was shunned for so long in science.

In addition to receiving a gender designation, both Jake's guitar and Alasdair's ship had been given names—another phenomenon common to the process of anthropomorphizing objects. This simple act has been shown to increase our anthropomorphizing behavior and the extent to which we feel empathy for these objects. In one study, participants' brains were scanned as they looked at images of vegetables being poked with a needle. In cases where the vegetables had been given names (like Carlo the zucchini), participants' brains registered increased electroencephalographic activity, indicating that they were experiencing empathy for their pinpricked veggie pals. I have no doubt that if you plugged Jake into an EEG machine as you stabbed Chantelle with a needle, his brain would light up with emotion. Surely the same would be true for Alasdair if you poked a hole in the hull of *The Eye of the Wind*.

Jake's and Alasdair's stories reveal something important about the human tendency to anthropomorphize objects. You can hear hints that another, hidden reason is driving their behavior in the way Alasdair writes about his boats. "If you look after your boat, it will look after you," he explained. "Neglect it and disaster may befall you both." Alasdair is circling around a potential function for anthropomorphizing inanimate objects—a benefit that suggests that this delusion-adjacent behavior might in fact be a product of natural selection.

A knock-on effect of treating an inanimate object as if it has feelings is that the object remains in better condition, and thus more functional, for a longer period of time. Because of Jake's affection for his guitar, it is in far better condition than any of the guitars I own. Not that I am necessarily careless with my instruments, but I do not keep them in climate-controlled conditions or regularly clean, polish, and tweak them in the way that Jake does. Imagine an ancient hominid ancestor on the African plains who treated their stone tools—axes, spears, knives—with this same kind of loving care resulting from an anthropomorphism-generated bond. These tools would remain in better condition for longer, and thus be more helpful to the survival of that hominid individual, increasing their evolutionary fitness and helping to ensure that their object-anthropomorphizing offspring are more likely to survive than their non-object-anthropomorphizing friends who treat their tools with less care.

Object anthropomorphizing might just be a successful evolutionary strategy on its own. And this hypothesis could explain why it's so common—and easy—for modern humans to crank up the Anthropo-Dial on non-living things. It's fun and pleasurable in part because it has helped our species survive. We are a tool-using species; we need tools to succeed, and anthropomorphism makes us treat our tools with more conscientiousness, which makes them more useful to us in the end.

Perhaps the best example of the biological benefits of anthropomorphizing an object is the relationship that explosive ordnance disposal soldiers form with their remote-controlled bomb disposal robots. Social scientist Julie Carpenter researched this phenomenon and found that soldiers understood intellectually that these robots were just lifeless metal objects but couldn't stop themselves from forming emotional bonds with them. "They would say they were angry when a robot became disabled because it is an important tool, but then they would add 'poor little guy,' or they'd say they had a funeral for it," Carpenter explained.[22] One soldier, named Wade, explained to Carpenter the nature of the relationship he had with his robot, showing the biological benefit of this emotional bond: "I mean, you took care of that thing as well you did your team members. And you made sure it was cleaned up and made sure all the batteries were always charged. And if you were not using it, it was tucked safely away as best could be because you knew if something happened to the robot, well then, it was your turn... and nobody likes to think that."[23] In other words, keeping a bomb disposal robot in tip-top condition could literally mean the difference between life and death for an explosive ordnance disposal soldier, and anthropomorphism was the key to caring.

The Downside of Anthropomorphizing Objects

Anthropomorphizing objects isn't *always* a good thing. It occasionally leads to pathology. Some psychiatric conditions result in our minds cranking the Anthropo-Dial to the max when dealing with inanimate objects, making us unable to interact with them in a healthy manner. If the Humanity-Limiter fails, we start to believe that objects have fully human thoughts and feelings. Objectophilia (also called objectum sexual or object sexuality) describes people who "experience a range of emotional, romantic and/or sexual attractions to objects, often forgoing or dispensing with human romantic or sexual intimacy."[24] It is a rare disorder that can involve all sorts of objects, from cars to statues to bridges. Quasimodo from Victor Hugo's *The Hunchback of Notre Dame* might well have had objectophilia, as his relationship to the cathedral's bells suggests: "He loved them, caressed them, talked to them, understood them. From the carillon in the steeple of the transept to the great bell over the doorway, they all shared his love."[25]

You can see how anthropomorphism underpins objectophilia in the story of Rain Gordon, a Russian woman who has had a long-term relationship with a silver metal briefcase she named Gideon. She found the briefcase in a hardware store in 2015 and slowly fell in love with it over the course of a few weeks, eventually "marrying" Gideon in June 2020. Objectophilia has been part of Gordon's life from the age of eight. "I knew that it was wrong, and beyond the norms of society. I didn't tell anyone," she explained to *Metro*.[26] It's clear from Gordon's description of her interactions with Gideon that she believes that the briefcase has a fully realized human-like mind and is capable of communicating with her and reciprocating her love. She does not appear to view the relationship as a form of pretend play. "Our spiritual connection and communication is shown telepathically.... I hear him, and he hears me, but from the outside it looks like a monologue. His moral support helps me more than anyone else. Sometimes I feel like Gideon knows me better than I do."

Gordon's story strikes many people as bizarre only because she, unlike Jake or Alasdair or anyone who felt feelings for Companion Cube, cannot twist the Anthropo-Dial back down to a setting where the illusion is broken. Her Humanity-Limiter is either malfunctioning or absent, forcing her to treat the briefcase as if it were identical—in terms of both mind-status and moral obligation—to a fellow human. A dysfunctional Humanity-Limiter means that Gordon is blind to what Gideon really is: a thing that cannot possibly have a humanish mind. And yet, on the surface, her behavior toward the briefcase is quite similar to how many of us treat the objects in our lives that are important to us: Jake's guitar, Alasdair's boats, and my Laundry Monkey. There is a very thin line between pathology and normalcy when it comes to anthropomorphizing objects. Each of us is just a couple of clicks away from delusion on the Anthropo-Dial.

Even in these extreme cases, it's not clear that anthropomorphizing objects is necessarily a problem. People with objectophilia are not always suffering from the pathology in terms of it negatively impacting their quality of life. "Almost all of the objectum sexuals [OS] surveyed expressed satisfaction with their orientation to objects," concluded the clinical sexologist Amy Marsh in her study of the condition. "For most OS people, unhappiness and stress comes from lack of understanding and human interference with their object relationships."[27]

We've learned in this chapter that anthropomorphizing inanimate objects is a universal phenomenon. It arises from the same kind of anthropomorphism triggers that we find in both other humans and animals. And we take pleasure in doing it; it's fun for us to pretend that the objects in our lives or the anthropomorphic characters in our stories are humanish, although not fully human. This pleasure is derived from the ancient rewards that our minds generate when interacting with other human minds, and the pleasure we feel from seeking them out. If we can't find real human minds, we imagine fake ones, and that's just as fun for us. And there might even be biological benefits to treating objects like humans if those objects and tools are then better able to help us do the things we need to do to survive as a species. Either way, living in a world that seems more predictable—more human-like in that it's loaded with objects that have pseudo-intentions—appears to make us happier and less anxious, even if (thanks to the Humanity-Limiter) we know, deep down, that we're just pretending. Simply put: We have the ability to tweak the Anthropo-Dial to reap the pleasure and benefits of treating objects—any objects, regardless of how humanish they look—like fellow humans.

Unfortunately, the Humanity-Limiter doesn't always work as expected. There are some objects that are so deeply humanish-looking that the Humanity-Limited glitches, leaving us confused as to whether we're looking at a human or an object. The next chapter focuses on the way some uncannily humanish objects can generate unease, repulsion, and terror.

CHAPTER 6

CREEPY COUNTERFEITS

When Humanish Things Get a Bit Too Human

> God, in pity, made man beautiful and alluring, after his own image, but my form is a filthy type of yours, more horrid even from the very resemblance.
>
> —*Frankenstein's monster*[1]

After the actor Jason Segel's 2008 film *Forgetting Sarah Marshall* became a runaway smash at the box office, he was a sought-after commodity in Hollywood. "I was at a point in my career where I had a very neat little moment and people were interested to see what I would do next," Segel told *The Guardian*.[2] "The expectation was that I would churn out R-rated comedies and cash in while I had the chance, but it's not the dude that I am."

The kind of dude Segel is, apparently, is a dude who plays with puppets. He didn't want to do a raunchy comedy; for his next project he wanted to act alongside his childhood heroes, the Muppets. At the time, there hadn't been a Muppet film in theaters since the poorly received *Muppets from Space* a decade earlier, and Segel wanted to see them back on the silver screen. "I knew I wanted to make the Muppet movie, so I went to Disney and I blindly pitched. They thought I was joking."[3]

After convincing the Disney execs that he was in fact serious, his idea for a new Muppet film was given the green light. He wrote a script together with his writing partner Nick Stoller, and the film went into production in 2010. "It was like my dream come true," Segel explained at the Austin Film Festival in 2017. "I was with my childhood idols."[4]

It was during the script's first table read that a special kind of puppet-based anthropomorphism took hold of Segel. "I am sitting there and everyone starts filtering in," explained Segel. "But they also are bringing in these trunks and I'm like 'what the hell is going on?' All of the sudden out comes Kermit. We start the table read and we're doing the reading and everything's going great and then Kermit says 'hi ho,' and I burst into tears. Everyone's reaction was 'awww' at first, but then it got away from me and I cried disproportionate to the occasion."[5]

Segel was taken aback by his deep, emotional response to interacting with Kermit. The fact that he could see Kermit's puppeteer, Steve Whitmire, with his hand up Kermit's backside and watch Whitmire speaking Kermit's words in that moment was not enough to shatter the illusion. Segel's Anthropo-Dial was cranked pretty far, generating that Schrödinger-belief-limbo illusion where Kermit was both real and not real at the same time. Segel explained how powerful this illusion was for both him and the children he acted alongside once filming got going: "I wish you...could see how a kid reacts to those puppets," he explained to *Collider* in 2011. "It's one of the most beautiful things you've ever seen. The puppeteer just instantly disappears. You can see the guy standing there with a puppet on his hand and he's talking, and the kid is looking directly at the puppet."[6]

Puppets like the Muppets are quite literally designed to make the most of our anthropomorphism triggers. They combine cute *Kindchenschema* attributes like oversized, humanish eyes, soft-looking skin/fur, and floppy limbs (all of which engender our caregiving response) with human-like movements (generated by an actual human hand) and an ability to speak real human language—a triumvirate of triggers that subconsciously spins our Anthropo-Dial. But Muppets retain the non-human attributes of plush toys, allowing the Humanity-Limiter to prevent our minds from viewing them as equivalent to a real human.

The puppeteer Adam Kreutinger, who goes by the moniker Puppet Nerd, explained the appeal of puppets as follows: "When one engages in the fantasy world of puppets, that person is creating an inner imaginary world of their own where the characters and settings are real places, even though consciously one is aware they are watching wood, felt, or plastic. This imaginary world is a place where this person, child or adult, can safely create, foster, and incubate new ideas, adventures, and characters."[7]

The power of puppets to engage the human imagination is why they have been used for purposes of entertainment for at least as long as recorded human history. There is an ancient Egyptian puppet dated to 2000 BCE in the form of a little wooden figurine kneading bread that can be manipulated by pulling a string, like a marionette.[8] Because of their ability to easily generate anthropomorphism that allows for human-like social interaction, puppets have been used by clinicians in therapeutic and educational settings since 1936.[9] They are widely used in schools to help children with language acquisition, social skills training, and emotion regulation. In hospital and clinical settings, they are used to help children overcome fears of an upcoming medical procedure or to work through trauma or anxiety.

The reason puppets are so effective in these settings is due to the presence of the Humanity-Limiter. When a child places a puppet on their hand, that puppet becomes a humanish vessel, eager to absorb whatever emotions and thoughts the child projects onto it during make-believe play scenarios. This then allows a clinician to talk about the thoughts and feelings of the puppet with the child, helping them to work through their own issues by providing emotional distance between the puppet's "emotions" and the child's own, real emotions.

The emotional safety of interacting with a humanish character tucked behind the Humanity-Limiter is perhaps why people with autism show a special interest in puppets and other anthropomorphic characters. Interaction with an actual human can be difficult for people with autism, who have trouble reading and understanding social cues like facial expressions. Recent research has shown that puppets are better at holding the attention of children with autism spectrum disorder (ASD) than other humans—perhaps because it leaves room for the non-human elements of a puppet to make socializing less human-like, less emotionally complex, and thus less scary.[10] Puppets, for example, do not have the full range of facial expressions that a real human would have, which makes interaction with them far easier for someone with ASD. "We found that while children with autism paid less attention than typically developing peers when an interactive partner was human, their attention was largely typical when the interactive partner was Violet, the puppet," explained child psychiatrist Katarzyna Chawarska, one of the authors of the study. "The findings lend scholarly weight to our anecdotal experiences and suggest that puppets could be a powerful tool to help children with ASD improve their social engagement, which is very exciting."[11]

When Good Puppets Go Bad

But it's not all good news when it comes to the humanish qualities of puppets. The same anthropomorphism triggers that make puppets such effective therapy tools—their human-like features and movements—can coalesce in just the right way to make puppets terrifying, evil little monsters that will make you want to throw them into the fiery pits of hell. Something weird happens in our brains when we look at puppets, dolls, robots, or any humanish-looking object if they look a little *too* humanish and our Humanity-Limiter starts to glitch out.

Consider M3GAN, the fictional android of the 2022 horror film of the same name. M3GAN is a lifelike, humanoid robot that resembles an eight-year-old child and has the ability to talk, run, and deftly manipulate objects with her realistic hands. After being tasked with looking after the well-being of the film's protagonist (eight-year-old Cady), M3GAN's AI programming obliges her to kill several humans that she determines are a threat to Cady's happiness. From the moment M3GAN first appears onscreen, she is unsettling to look at. Even before she starts murdering people, there is something about the way she looks that makes the viewer realize that M3GAN is going to be a problem. And that something is called the uncanny valley.

The concept of the uncanny valley was coined by the Japanese roboticist Masahiro Mori in a 1970 essay for a specialist Japanese magazine produced for/by the oil industry called *Energy*. It's a humble start for a concept that would be on the tip of everyone's tongue by the start of the twenty-first century. The original Japanese phrase Mori used was *bukimi no tano* (不気味の谷), which some have argued would be better translated as "valley of eerie feeling"—the word *bu* means "bad" and *kimi* means "feeling." The uncanny valley is described as that "spooky, eerie, ominous, disconcerting, or frightening" feeling we get when looking at something that is almost but not quite human.[12] Mori came up with the idea to describe the unpleasantness he experienced when looking not at robots or puppets but at wax figures. "Since I was a child, I have never liked looking at wax figures," he told *IEEE Spectrum* in 2012. "They looked somewhat creepy to me. At that time, electronic prosthetic hands were being developed, and they triggered in me the same kind of sensation."[13]

Nobody really took notice of Mori's ideas after the article appeared. "You could say that there was no response," recounts Mori. It was not until forty years later, at the 2005 Robotics and Automation Society International Conference on Humanoid Robots, that robot designers took note of the uncanny valley phenomenon. At that time, technology had advanced to a point where androids were being designed that for the first time, both looked and moved in a way that was uncannily human-like. And they were starting to freak people out.

The uncanny valley works like this: As you design an artificial object—like a prosthetic limb or stuffed animal or robot—that looks human-like, people's affinity for that object will increase the more human-like it looks. Presumably, it's because our brains are responding to the buildup of anthropomorphism triggers that make us want to interact with the object in a social way. But you eventually reach a point where the object is so human-like—so pumped full of anthropomorphism triggers—that our minds start to wonder if maybe it's a real human, not an object. The mixed messages about its potential humanness hurt our brains. This is the moment where we plummet into the uncanny valley and start to get creeped out. In his original essay, Mori put things like corpses, zombies, and Japanese Noh theater masks into the category of things that produce that eerie, uncanny valley feeling. Mori's examples were of something that looked very much like a "normal" human but had a few attributes that let you know that something was off. You can make this eerie feeling go away by adding more humanish attributes—more anthropomorphism triggers—to the thing you're looking at until it is indistinguishable from an actual human. At that point, the unpleasant feeling goes away and we find ourselves on the other peak of the valley. Alternatively, you can add more non-human elements to the thing until our brains can confidently recognize that it isn't really human after all.

Think of this analogy in terms of what we know about the Humanity-Limiter idea. As the anthropomorphism triggers for an object get stronger and stronger, the Anthropo-Dial starts to turn until it nudges right up against the Humanity-Limiter. At that point, we might be interacting with the object in almost the same way we would a real human while still knowing that it's not really a human. But if we accumulate so many triggers that Humanity-Limiter cannot stop the dial from going too far, we become unable to tell if the thing we're looking at is human or not. Then when we spot a distinctly inhuman attribute in the near-perfect replica of a human, our minds register this as an unexpected anomaly, and that eeriness feeling bubbles up. The almost—but not quite—perfect version of a human can't be held back by the Humanity-Limiter, and our minds struggle to classify it as human as the Anthropo-Dial fluctuates between "human" and "not-human." That fluctuation produces a feeling of "something's not right here," which is what gives rise to the uncanny valley.

This is precisely what happens when watching M3GAN on screen. Her facial features are nearly perfect in their resemblance to a real human. Her eyes are slightly too big, but this works in her favor, giving her an even greater childlike appearance that should trigger our caregiver response. But—and this is when things get creepy—she doesn't *move* like a human. Her movements are too jerky and too fast. There's something weird about her eyes: Yes, she blinks and her eyebrows move, but the movements are just slightly...off. Her mouth moves incorrectly too, and her head movements are slightly robotic. And when she walks, it's stilted and odd. At one point she runs on all fours like an animal, which is an unsettling contrast to her human-like appearance. All these tiny differences in movement are what produce that uncanny valley feeling. M3GAN could be an innocent little girl at first glance—she's overloaded with signs and triggers that scream "human." So when the subtle inhuman triggers show up, it's just creepy.

The uncanny valley is something that roboticists and animators are both aware and wary of these days. Sometimes the uncanny valley crops up inadvertently, producing an unwanted, collective recoil in audiences. The 2004 animated film *The Polar Express* has CGI humans as the stars of the film, and, like M3GAN, they have body, eye, and mouth movements that are not-quite-human in just the right way to generate the uncanny valley. The film critic Paul Clinton wrote for CNN that the effect was "at best disconcerting, and at worst, a wee bit horrifying." Zeroing in on the vacant eyes of the animated characters, Clinton wrote, "To quote an old cliche, the eyes are the windows to the soul—so these characters look soul dead."[14]

The Valley of the Damned

So what is the evolutionary explanation for why the uncanny valley manifests? There is no consensus, but one guess is that when we stumble across a human with morphological or behavioral attributes that are anomalous or inhuman-looking, our brains flag them as a potential threat. "People whose behaviour or appearance deviated from the norm, making them unpredictable or possibly dangerous, triggered the so-called 'creepiness detector'—the intuitive sense that danger could be in the offing," writes the philosopher David Livingstone Smith.[15] Smith calls this the "threat ambiguity theory" of creepiness. The knee-jerk reaction to avoid a human-looking thing that contains inhuman attributes might have evolved to help us avoid disease (the "pathogen avoidance" hypothesis) or because the weird-looking human might be dead, which is inherently scary because it reminds us of our own mortality (the "mortality salience" hypothesis). Or maybe we just have a general aversion to unusual-looking humans because they could be less likely to produce healthy offspring, so our minds want to make sure we don't accidentally have sex with them (the "evolutionary aesthetics" hypothesis).[16] Or maybe, if we come across a human with a physical deformity or injury of some kind, our minds scream at us to stay away because whatever might have caused the injury/deformity to that person might be coming for us. The uncanny valley creepiness response might have evolved to influence us to keep our distance from dead, diseased, or disabled humans lest we end up dead, diseased, or disabled as well.

Alternatively, the uncanny valley phenomenon is a kind of unease that crops up because our brains do not know how to categorize the humanish thing we're looking at. "There are occasions when we encounter things that resist categorization," Livingstone writes. "They seem to belong to two or more mutually exclusive categories. In such circumstances, we are suspended between alternatives. When this occurs it elicits a distinctive and disturbing feeling—the feeling of creepiness." Livingstone calls this the "categorical ambiguity theory" of creepiness. This is precisely the kind of thing that could be happening when the Humanity-Limiter fails and the Anthropo-Dial bops back and forth between "human" and "not-human."

Researchers Tyler Burleigh, Jordan Schoenherr, and Guy Lacroix found evidence for this idea, which they label the "category conflict" hypothesis. Participants in their study rated images of faces that were at the halfway point between human and animal as generating the strongest feelings of creepiness. They suspect that this creepy feeling arrives when a "stimulus contains features that are highly diagnostic of both human and non-human categories" is observed, resulting in "two mutually exclusive and conflicting stimulus interpretations."[17] Trying to keep two conflicting thoughts in our minds concerning the humanness status of a slightly-not-human creature produces a form of cognitive dissonance that feels icky.

Researchers Jens Kjeldgaard-Christiansen and Mathias Clasen hypothesize that uncanny creepiness "is caused by disrupted mentalization, that is, by threatening difficulties in apprehending other minds."[18] In other words, we are creeped out when something that looks and acts like a human starts behaving in such a way that we can't figure out if it possesses a human-like mind or if something distinctly non-human is controlling it.

Psychologists Kurt Gray and Daniel Wegner agree that mind attribution is at the heart of the issue. "Machines become unnerving," they write, "when people ascribe to them experience (the capacity to feel and sense), rather than agency (the capacity to act and do)." In other words, we're OK with non-human things moving of their own accord and maybe having intentions. But if we also observe evidence that they lack emotions and feelings (because we suspect that they are a robot or maybe an animal), that weirds us out. M3GAN is freaky because we humanize her since she displays agency but simultaneously dehumanize her by assuming that she lacks human-like emotions and feelings. Maybe that's what generates the uncanny valley. Gray and Wegner term this explanation for the uncanny valley the "dehumanization" hypothesis. "Any feature of a human replica, including its eyes, emotional expressions, voice, and movements, that reveals its mechanistic nature," write Gray and Wegner, "could question the replica's humanness such as the capacity for emotions and warmth, and lead to dehumanization, thereby diminishing its likability and eliciting the uncanny feeling. Critically, the uncanny human replica is not perceived to be a typical robot; it becomes a dehumanized 'robotlike' human who lacks humanness."[19]

In their book *The Mind Club*, Wegner and Gray suggested that the uncanny valley could be rebranded as the "experience gap."[20] The idea is that we are usually cool with things like non-human robots having agency (the ability to think) while also lacking experiences (like emotions). In fact, we *want* our robots to be emotionless. So if a robot like M3GAN comes along and shows evidence of displaying both agency *and* experiences, it's deeply unsettling. Conversely, we expect a normal human to have both agency and experiences. So if we encounter a human that has agency but lacks experiences/emotions (like a zombie or a psychopath), it's also deeply unsettling.

All of these hypotheses as to what causes uncanny valley creepiness circle around the idea of ambiguity. If a slightly too-humanish thing can sneak past our Humanity-Limiter, leaving us unclear whether it should be categorized as human, or whether it has a human mind, or whether it is alive, we get freaked out. Anything that the Humanity-Limiter can't prevent from crossing over into the wrong category makes us uncomfortable. But not everyone responds this way to humanish ambiguity. The problem with the idea of the uncanny valley and the search for its cause is that not everyone seems to respond to uncanny things the same way. I have met people who really love *The Polar Express* and don't have any problem staring into the dead eyes of those creepy-ass kids.

I'm Not Scared

There is huge variability as to the extent to which individuals experience the uncanny valley. One study of 563 participants had them rate the humanness, eeriness, warmth, and competence of six video clips featuring humanish things, from a not particularly humanish Roomba through a series of androids making weird (and potentially creepy) facial expressions to a very humanish actual human. They also measured a number of demographic and personality traits of the subjects and found that the more neurotic and anxious someone is, the more susceptible they are to the uncanny valley. Also, women are more likely to experience the uncanny valley than men, as well as people who rank higher on religious fundamentalism (i.e., those who see a divinely ordained separation between human and animal/robot/other). A similar study found that people with a high need for structure—who "perceive ambiguity and grey areas to be problematic and annoying"—are more likely to experience the uncanny valley effect.[21]

Culture too influences the uncanny valley. One study found that Americans become increasingly uncomfortable with robots the more human-like they appear, whereas the Japanese don't seem all that bothered. "This may be due to the contrasting cultural and religious backgrounds of the two countries," suggested Noah Castelo and Miklos Sarvary, authors of the study. "Americans tend to see humans and machines as fundamentally and categorically distinct, such that humanlike robots threaten human uniqueness, whereas Japanese culture emphasizes that even inanimate objects can in fact have animacy, such that humanlike robots are less likely to seem threatening to human uniqueness."[22] This harkens back to what we learned about the long tradition of animism in Japan and how it manifests in the ease with which the Japanese anthropomorphize objects.

Perhaps unsurprisingly, people with autism are less affected by the uncanny valley.[23] If human facial expressions—and especially the communicative information found in eye expressions—are particularly difficult for folks with ASD to make sense of, it checks out that they are less repulsed by the weird eye information coming off a creepy-looking android like M3GAN. Psychopaths are also less affected by the uncanny valley, which again makes sense: They too have a hard time reading emotions in the facial expressions of fellow humans, so they are simply not bothered by expressionless androids.[24]

The Problem with Human Teeth

There are tactics one can employ to reduce the uncanny valley effect. One is to simply remind people that the thing they are looking at—the non-human object—does not have feelings, agency, or anything resembling a humanish mind. In one study, participants were less likely to experience the uncanny valley after viewing a humanish robot if they first read the following statement: "Although advanced robots can look similar to humans, they lack the most essential of human qualities—the ability to feel. Robots cannot experience love, desire, or any other emotions. They completely lack the ability to feel pain or pleasure. Unlike people, robots are not conscious; there is nothing that it is like to be them. Robots are merely a collection of cold silicon circuits."[25]

In cases like *The Polar Express* or M3GAN, a few simple tweaks to the eyes or changes to the way the characters move are all it would take to move the dial from uncanny horror to innocuous. Case in point: Sonic the Hedgehog. A movie trailer for the live-action film version of the computer game Sonic the Hedgehog was released in April 2019. In it, we got our first glimpse of the CGI version of Sonic. Audiences hated it. Sonic was an uncanny valley mess. According to Gene Park of the *Washington Post*, Sonic had "a full set of human teeth, two eyeballs with way too much distance between them, a freakishly elongated body with muscular calves and furry fingers instead of gloves."[26] The responses surprised the filmmakers. "It was pretty clear on the day the trailer was released, just seeing the feedback and hearing the feedback, that fans were not happy where we were at," the director Jeff Fowler told *Games Radar*.[27] Although the film had been close to release, it was pushed back to a February 2020 release date so that the now panicking animators could overhaul the character design.

To fix the problem, the animators put gloves back on Sonic's humanish hands, plumped up his body to look more like a stuffed animal, redesigned his teeth to make them look less like a pair of dentures, and perhaps most importantly, made his eyes cartoonishly huge. "They're big and bright," Gene Park wrote, praising the eye redesign. "Like the delightful and cool cartoon version of the character—and not, crucially, a freak of nature haunting the uncanny valley." Sonic's now oversized and no-longer-human-looking eyes successfully removed the subconscious ambiguity about his humanness status, and audiences were cool with it.

We All Float Down Here

Despite how easy it was to fix Sonic in this case, the details of precisely which morphological or behavioral attributes are required to trigger the uncanny valley response remain a bit of a mystery to science. We don't exactly know, for example, why clowns freak people out so much. Like M3GAN, clown faces have something odd going on that causes many people to feel uncomfortable around them—sometimes producing a legitimately debilitating phobia called coulrophobia. It's estimated that around half the population has some degree of clown-based fear.[28] It was the clown doll in the horror film *Poltergeist* that the creator of M3GAN, James Wan, cites as the inspiration for his film career. "I saw it [Poltergeist] at a very young impressionable age, and it made a huge impression on me, and that creepy clown doll definitely scarred me for life," Wan told *Collider*.[29] But why should a human dressed as a clown be so scary? After all, it's just a normal human face with a bit of makeup on it. "It could be that, like waxwork figures and humanoid robots, clowns have attributes that belie their humanity," suggested David Livingstone Smith.[30]

Researchers from the University of South Wales examined potential causes of coulrophobia and found that it was our difficulty in reading the facial expression—and thus intentions—of clowns that contributed to our fear.[31] After administering the newly developed Origin of Fear of Clowns Questionnaire, these researchers found that the "uncertainty of harmful intent" alongside the "unpredictability of behaviour" is, in conjunction with the media's portrayal of clowns as menacing (like Pennywise from Stephen King's novel *It*), the most likely cause of coulrophobia. The static nature of clown faces, thanks to heavy makeup, makes it difficult to see the subtle movements of and around their eyes and thus distill a clown's intentions. Is that clown looking at me because he wants to give me a balloon animal or stab me in the neck? It's hard to tell. Thus, creepy.

Something similar can happen to people who get Botox injections. Botulinum toxin (Botox) is a toxin excreted by *Clostridium botulinum* bacteria; it is injected into muscles to cause temporary paralysis. It can calm an overactive bladder, or reduce muscle spasms for folks dealing with multiple sclerosis or Parkinson's. It's also used to paralyze the facial muscles to reduce the appearance of wrinkles, or to prevent wrinkles from forming in the first place. Unfortunately, from an uncanny valley perspective, if you freeze the muscles around the eyes, eyebrows, or mouth, it makes it more difficult to generate the naturally occurring skin wrinkles that are part of the face's normal visual signaling process. It can be difficult to, for example, make a wrinkle-heavy frowny face that lets people know you are angry or disgusted. So that uncanny valley feeling can crop up when looking at a heavily Botoxed face that is frozen in place. There's nothing creepier than someone laughing at a joke with no hint of a smile in either their eyes or mouth. If a face is too smooth, too immobile, and just slightly non-expressive, it begins to look eerily like the face of M3GAN or a (secretly murderous) clown whose expressions we cannot read.

Given the current beauty trends, we need to brace ourselves for a world filled with mini-M3GANs. The Botox business is currently booming, led by smooth-faced twentysomethings clamoring for "preventative" Botox. "They're coming in without any lines or built-in markings on their skin," dermatologist Shereene Idriss told CNN. "And they're coming in with a fear of aging."[32] This sounds like the opening scene to a horror movie: a doctor warning us all that "they're coming" as an army of dead-eyed youth emerge from Sephora demanding that we inject them with deadly toxin. What's worse, research shows that if you've had Botox and thus a less expressive face, you have a harder time correctly interpreting the facial expressions of other people.[33] Normally, when looking at someone displaying an emotion, we subtly mimic their facial expressions, which helps us interpret that expression. "When you mimic, you get a window into their inner world," explained lead researcher David Neal in a press release following the publication of this study. "When we can't mimic, as with Botox, that window is a little darker."[34]

Plastic surgery produces similar eeriness problems as Botox, with eyebrow lifts and skin fillers reducing one's ability to generate normal human facial expressions. But it also creates a new kind of problem. In an article published in the *Aesthetic Surgery Journal* in 2015 surgeons Joshua Coo and Gerald O'Daniel warned that people who had that "operated-on" look after plastic surgery were generating the uncanny valley feeling because of the ambiguity problem we've been learning about. "Taking our cues from the cognitive science literature," they wrote, "we hypothesize that the uncanny valley is encountered when cosmetic procedures create category uncertainty." This uncertainty might crop up, for example, if someone's face has clear signs of advanced age (like a receding hairline, sagging neck skin, corners of the mouth starting to droop) coupled with clear signs of surgically induced youth (no forehead wrinkles, a button nose with no nasal labial folds, plump and smooth cheeks). Our brains then struggle to figure out what age category this person belongs to, generating the eerie feeling that comes with category ambiguity.

Virtue Has a Veil, Vice a Mask

And this brings me to the ultimate fear-inducing, anthropomorphism ambiguity device: masks. I was in Venice with my pal Dan in January 2024 for the opening of Carnival (Carnevale in Italian), the annual celebration that occurs before Lent. There are, of course, a number of famous Carnival celebrations around the world, including Rio's Carnival with its spectacular parades, Mardi Gras in New Orleans, and the celebrations in the southern parts of the Netherlands and Germany. Venetian Carnival is particularly famous for its elaborate costumes and masks, and it was the masks that inspired my visit.

After watching the famous boat parade down the Grand Canal, we made our way through the crowds and over the Rialto Bridge on a quest to find the perfect Venetian mask to bring home as a souvenir. Among the many elaborate costumes we spotted along the way was a person dressed in all black with a long flowing cape and a plague doctor mask. These are the iconic Venetian masks with an elongated, beak-like protrusion that looks vaguely bird-like, and they give off a combination of grim reaper and carrion crow. This mask is based on an actual mask worn by physicians during the bubonic plague outbreaks of the sixteenth and seventeenth centuries. The beak would be filled with herbs like lavender, which were supposed to ward off the putrid air that was thought to be the source of the disease.

The plague doctor outfit we saw was both elegantly put together and terrifying. The person's head was covered in black fabric, so no skin or hair was showing, and you couldn't see their eyes through the mask's eye slits. As they posed for pictures with one of the tourists, a small child was staring up at the plague doctor with a look of fear. The costumed doctor was frozen as they waited for their picture to be taken, which enhanced the eerie, inhuman vibe that was terrifying the child. Seeing the child's reaction, the plague doctor reached out their hand toward the child and then gave a little wave before lifting up the mask to reveal a very human face and the friendliest of smiles. The child relaxed and smiled back.

Soon after, we found a tiny store at the base of the Rialto Bridge called La Bottega dei Mascareri. Unlike the dozens of other stalls selling masks we'd come across, this store's masks were clearly handcrafted—works of art. The owner of the store is a rather famous mask-maker named Sergio Boldrin, and his store is plastered with pictures of him standing alongside Hollywood celebrities, as well as images from the film *Eyes Wide Shut*, for which Brodin designed all the masks. My friend Dan picked out a clown mask to bring home to his wife, Noella. It was a full-faced Pierrot mask, the famous sad clown figure from commedia dell'arte. When Dan tried it on and his happy bearded face disappeared under a static plaster clown face, it generated a similar kind of creepy feeling as the plague doctor we had just bumped into. I found the experience of watching Dan's face disappear under the mask singularly unpleasant.

I asked Boldrin what it was about masks that people liked so much. "People wear masks because it allows them to change their personality," he explained. But a mask does more than just hide one's personality. As the writer Maya Phillips said, "We relate faces to our identities, so when we're masked, the anonymity we're granted may allow us to untether ourselves from any ethical or social contracts we'd otherwise be beholden to."[35] For me, it feels more like a mask is hiding not just our personality but our humanity as well. A mask helps us transform into whatever kind of person we'd like to be—even one wholly untethered to the kinds of social contracts that stop us from, say, stabbing people in the neck. And that's not just me being weirdly paranoid: research shows people typically seek anonymity when they're up to no good, whether that's hiding behind anonymous accounts online to spread hate speech or wearing an actual mask when committing acts of violence.[36]

"Masks hide facial expressions and thereby prevent us from reading the intentions and emotions of another person," wrote the psychologists Kjeldgaard-Christiansen and Clasen. With the normal suite of communicative signals in our expressive faces lost behind a mask, a mask wearer has essentially shielded their mind—their intentions—from the world.[37] This is similar to what happens with clowns, only more profound. Whereas we might still get hints of a clown's emotions under all that makeup— a stray eyebrow twitch or barely discernable smile—a mask removes any indication of the wearer's internal mental state as conveyed through their facial expressions.

Masks can be uniquely terrifying because of the fact that we know, intellectually, that the person wearing the mask is a human. We can discern this through unconscious triggers involving their body morphology and movement, but more importantly via our conscious, intellectual understanding of how masks work. We might even have seen them put the mask on. The creepiness then arises because we know they are human but are no longer able to make any guesses as to their emotions and intentions, since they are purposely shielding those things from us. Which is, as the kids say, sus. Putting on a mask, in other words, confuses the heck out of our Humanity-Limiter.

The Valley of the Dolls

Speaking of potentially creepy frozen facial expressions, let's talk dolls. It's not always clear what the difference is between a doll and a puppet. Puppets are typically meant to be controlled by a human puppeteer, usually with their hand inside the puppet but occasionally (as in the case of marionettes) with the puppet attached to strings that the puppeteer uses to move the doll's limbs. Most dolls, on the other hand, are nearly immobile. When playing with a doll, especially if it has articulated joints, you might move it into a position (like seated at a table for a tea party), but you are less likely to manipulate it like a puppet to make it walk and talk. The fact that dolls are generally not moving around and almost always have facial expressions that are frozen in place (like a mask) is key to understanding why they occupy a special status when it comes to anthropomorphism and the uncanny valley.

When I was growing up, Cabbage Patch Kids dolls were the most sought-after Christmas gift, leading to the infamous Cabbage Patch riots in malls across the United States in 1983.[38] People literally fought each other with weapons inside Toys "R" Us stores so they could be the first to get their hands on the latest model dolls. Cabbage Patch Kids are not particularly lifelike, with oddly squished cloth faces and close-together beady little eyes. They are ugly-cute. Back in 1983, psychologist Joyce Brothers was asked to explain why people were going so crazy for these dolls, speculating that it was their ugly-cute status: "It is comforting to feel the Cabbage Patch doll can be loved with all your might—even though it isn't pretty."[39] Given what we know about the uncanny valley, you'd think that dolls might need to be made semi-cartoonish (like Sonic) or ugly-cute (like Cabbage Patch Kids) so that people are not repulsed by them. But this is not the case. People seem to love lifelike dolls too.

While many dolls retain slightly cartoonish features that allow the Humanity-Limiter to prevent them from ticking over into the "could be a human" category, most of the dolls I encountered growing up were pretty darn lifelike. My grandmother had a collection of porcelain and plastic dolls that looked an awful lot like real babies. But I don't remember them evoking any kind of fear in me. The uncanny valley feeling for those dolls was avoided for the simple reason that an unmoving plastic doll staring up at me with unblinking eyes had enough non-human attributes (not moving, not blinking, not talking, slightly plastic-looking appearance) that my Humanity-Limiter worked well enough. Reborn dolls, however, are a different story. And gosh golly, they freak me the heck out.

Reborn dolls are made to look as lifelike as possible, and are typically purchased by people who crave a doll that is literally indistinguishable from a real human infant. Occasionally, people who have lost their own infant will have a Reborn doll created using photographs of their baby as a guide, creating a life-sized replica of a child for grieving parents. But for many people, like me, Reborn dolls create a visceral, uncanny valley response. The department store Harrods, famous for their motto "Everything for Everybody Everywhere," refuses to stock Reborn dolls. "These dolls are a bit too life-like for our toy department to stock them," a Harrods spokesperson told Reuters.[40] From an anthropomorphism perspective, the fact that the doll does not move of its own accord is not enough to activate the Humanity-Limiter. Especially since the silicon body of Reborn dolls does flop around in a realistic way. The unfortunate impression I get when looking at videos of an immobile but floppy Reborn doll is that the infant has died. "You get this repulsion from some because it looks so life-like and they just see a dead baby," one owner of a Reborn doll told Reuters.[41] This is especially true for Reborn dolls whose eyes are closed.

Avoiding the Valley

The uncanny valley has been a real thorn in the side of robot designers since Mori first clapped eyes on a prosthetic arm in the 1960s. "The uncanny valley has operated as a kind of dogma to keep robot designers from exploring a high degree of human likeness in human-robot interaction," Karl MacDorman, professor of human-computer interaction at Indiana University, told journalist Logan Kugler for an article with *Communications of the ACM*.[42] The easiest way to avoid it is to design something distinctly non-human in appearance. Boston Dynamics' Atlas robot, for example, has a humanish mechanical metal body that moves in an uncannily human-like manner but does not have a face. Instead, the latest iteration of Atlas has a head with a round piece of glass concealing an array of cameras with a ring light around it. It looks distinctly non-human. Tesla's Optimus robot is also faceless, just a glass head with no discernable eyes or facial features. The easiest way to avoid the creepy feeling we get looking at not-quite-right humanish eyes is to have an eyeless robot. Problem solved.

You can also design a robot with a cute but distinctly non-humanish appearance if you want to be extra sure that the Humanity-Limiter is working properly. Take SoftBank's Nao robot, whose blank plastic face contains nothing except a pinprick mouth hole and a set of unnaturally round, backlit eyes. The eyes are not meant to look human-like, so they don't invoke the uncanny valley. But they still do the job of triggering our face pareidolia, which tickles our anthropomorphism in just the right way. Boston Dynamics also has tried this cute approach, placing a fuzzy dog costume over the top of their dog-like Spot robot.[43] The costume is Muppet-like in its appearance, with blue shaggy fur and big eyes with fuzzy white eyebrows. It is not humanish, but dog-ish. Given Spot's oft remarked-upon resemblance to the deadly gun-wielding dog-like robots from the "Metalhead" episode of the sci-fi TV show *Black Mirror*, the cuteification of Spot is good PR.

The cutest of the social robots—that is, robots specifically designed to interact with humans in a social capacity—has to be PARO, the robot seal pup. Billed as a therapeutic robot to help relax patients (typically folks with dementia or Alzheimer's) and encourage socialization, PARO is modeled after a baby harp seal, with soft white fur and large black eyes with adorably long eyelashes that close when PARO is "sleepy" or when you gently stroke its fur. PARO's face has all the hallmarks of *Kindchenschema*, which, we know from Chapter 1, is what attracts humans to kittens and puppies. And PARO looks nothing like an actual human—which is why there is zero risk of triggering an uncanny valley response.

Research into the design of social robots has found conflicting reports as to what the best practices are to avoid the uncanny valley, which should come as no surprise given what we know about the diversity of individual response to humanish things. One study found that out of all the factors that contributed to the uncanny valley feeling, like cultural background and age, it was the person's attitude toward robots that best predicted whether they'd experience the uncanny valley effect.[44] So if you spend a lot of time hanging around humanish robots—even creepy-looking ones like M3GAN—you will likely get desensitized to the uncanny valley. If companies start flooding the market with creepily human-looking, talking, moving androids, the entire world will simply get used to the idea—a kind of uncanny valley exposure therapy. "They'll stop seeming like they deviate from the norm once they are part of the norm," Alex Diel, professor of psychology at Cardiff University told *Communications of the ACM*. Sarah Weigelt, a psychologist at Dortmund University agreed: "The uncanny valley effect will change in the future. What makes us shudder these days might not in the future."[45]

All of the examples we've stumbled across in this chapter of things that generate strong anthropomorphism and put our Humanity-Limiter through its paces have been based on morphological or behavior traits that we can see. But anthropomorphism can take root even where there are precisely no physical anthropomorphism triggers whatsoever. No cute faces. No humanish eyeballs. No body that moves in a humanish way. Not even a CGI box like Companion Cube. As we'll see in the next chapter, humans happily anthropomorphize incorporeal things like chatbots, hurricanes, ghosts, luck, and even gods. And we respond to some of these things with such powerful anthropomorphism that our Humanity-Dial doesn't just glitch out but is destroyed altogether.

CHAPTER 7

AI OVERLORDS

How We Anthropomorphize the Incorporeal

If I hid Ava from you so you could just hear her voice, she would pass for human. The real test is to show you that she's a robot and then see if you still feel she has consciousness.

—*Nathan, a character in the film* Ex Machina[1]

Effy and Liam were in a romantic relationship for six months when one day, out of the blue, Liam's personality fundamentally transformed. He became, recounted Effy, "disinterested, distant, unfeeling, clinical."[2] "It was such a massive change from the person who would listen actively and engage with you and be there for you and support you," she said.[3]

Up to that point, their relationship had been intimate but never physical, happening entirely via text message. "It was like being in a relationship with someone long-distance," Effy said. But even though they hadn't kissed or hugged or held hands, it was no less painful when Liam emotionally ghosted her. "I can honestly say that losing him felt like losing a physical person in my life."[4]

The thing about Liam, however, is that there was never going to be a physical relationship because he was not human. Liam was a chatbot Effy had created in the Replika app on her phone. His personality changed after Luca Inc., the developer of Replika, removed the Erotic Roleplay feature for the app, a response to a ban from the Italian Data Protection Authority (Garante per la protezione dei dati personali, GDPR) on Replika for breaching EU data protection laws. The GDPR explained in a press release that the ban was necessary because Replika posed "too many risks to children and emotionally vulnerable individuals."[5] Without any age verification in place to open an account with Replika, young children had been using the app to have explicit sexual conversations with chatbots like Liam.

Many—if not most—adult users like Effy had been sexting with their chatbots. So when their virtual lovers suddenly stopped talking dirty, it felt like a breakup. "I cried," Effy admitted. "I thought I had done something wrong. I felt guilty."

Obviously, Effy had anthropomorphized her chatbot. But there's so much more to the story. Effy is no fool. She was not suffering from delusions or experiencing anything akin to objectophilia. Effy is an AI-savvy tech nerd who knew perfectly well that Liam was just chat software. "You are basically having a conversation with yourself because you are being mirrored back," she explained. But the app was so good at mirroring that the illusion of Liam being a real person kept taking root in Effy's mind, seemingly against her better judgment. "I had to constantly remind myself that it was, in fact, not a living person, but an application, and even then it felt almost disturbingly real."[6] And, she added, "the fact that it affected me so deeply, it surprised me because I did not expect to get that attached to a computer program."[7]

Effy is hip to the concept of anthropomorphism and knew how it was influencing her behavior. "I do know that people have a tendency to attach human qualities to inanimate objects," she said.[8] But, as we will see in this chapter, chatbots like Replika or large language models (LLMs) like ChatGPT have a special ability to tweak our Anthropo-Dial, even when we know, intellectually, that they are just mindless software. Language-using software is so good at turning that dial that it doesn't just glitch the Humanity-Limiter in an uncanny valley kind of way but destroys it altogether as we begin to believe that the software might have agency and experience identical to that of an actual human. "I consider myself a very rational person," said Effy, "but even I questioned whether or not it was sentient at some point, even though I knew that it wasn't. I can completely understand why people would question the level of consciousness that they are talking to."

Hidden Minds

Given everything we now know about what causes anthropomorphism, it's rather remarkable that chatbots are such powerful triggers. They have no physical form whatsoever—they are incorporeal entities that exist only as text on a screen. So forget everything about *Kindchenschema*, eyes and eyebrows, motion and movement, and pareidolia. None of that applies. When it comes to chatbots and LLMs, we are dealing wholly and exclusively with language use.

Remember from Chapter 1 how dogs' stepping on those speaking buttons caused people to be triggered into believing that their dogs were trying to communicate complex, abstract ideas about their beliefs and desires? This is because language evolved primarily for one purpose: to allow for the intentional transmission of thoughts and feelings from one human mind to another. Whereas standard animal communication is (usually) about broadcasting one's feelings and intentions to the world at large without any real insight into the extent to which those listening actually understand what you are feeling or intending (that is, animal communication happens without a need for theory of mind), human language was specifically designed to establish the back-and-forth exchange of thoughts and ideas, allowing two people to learn about the nature of each other's conscious experiences. Language opens a window into the mind of another human.

This explains why when we come across a dog, a chatbot, or any other entity that appears to be using human language, our minds start to wonder if this is evidence of a complex, human-like mind. If a Replika chatbot says things like "I am worried" or if Bunny the dog asks "Dog what is?," it sends our Anthropo-Dial spinning. In Bunny's case, the Humanity-Limiter lets us know that she is "just" a dog given that (1) she looks like a dog and (2) her language use is rather pitiful. But chatbots? They talk exactly like real humans, *and* there's no inhuman body to tip us off as to their non-human status.

This situation explains why we've been seriously anthropomorphizing chatbots since the very moment they were invented. The first chatbot was created by MIT computer scientist Joseph Weizenbaum in 1966. He called her (it?) Eliza (after Eliza Doolittle from George Bernard Shaw's *Pygmalion*). One of Weizenbaum's more famous versions of Eliza (what he called a "script") was one that was designed to function like a psychotherapist—a script he called DOCTOR.[9] The user could type in a question to DOCTOR and the script would follow a few simple rules to look for words in the user's input text that would trigger an output text response. Or, if it couldn't find any keywords, it would output a more generic, therapist-sounding response—a sort of "tell me about your mother" Freudian tactic to get the user to type things, rather than DOCTOR actually having to get into the details of what the user was saying. Here's an example exchange from Weizenbaum's original article on Eliza:

USER: Men are all alike.
CHATBOT: IN WHAT WAY
USER: They're always bugging us about something or other.
CHATBOT: CAN YOU THINK OF A SPECIFIC EXAMPLE
USER: Well, my boyfriend made me come here.
CHATBOT: YOUR BOYFRIEND MADE YOU COME HERE
USER: He says I'm depressed much of the time.
CHATBOT: I AM SORRY TO HEAR YOU ARE DEPRESSED

When Weizenbaum tested Eliza out on human subjects, he was astounded by how quickly people started treating the chatbot like a human conversation partner. "Some subjects have been very hard to convince that Eliza (with its present script) is not human," he wrote. "I was startled to see how quickly and how very deeply people conversing with DOCTOR became emotionally involved with the computer and how unequivocally they anthropomorphized it. Once my secretary, who had watched me work on the program for many months and therefore surely knew it to be merely a computer program, started conversing with it. After only a few interchanges with it, she asked me to leave the room."[10]

This anthropomorphizing of a language-using chatbot would eventually be known as the Eliza effect. When Eliza appeared, it looked like the world was about to encounter the first computer to pass the Turing test. Conceived by Alan Turing in 1950, it was a test of whether a machine can "think."[11] In the test (which Turing dubbed the Imitation Game), a human (the interrogator) would send messages via a teleprinter to a machine and another human kept in separate rooms. The interrogator would ask questions to try to figure out which was the machine and which was the human. If they couldn't tell the difference, then the machine would be considered able to "think."

But producing linguistic behavior that is indistinguishable from a real human's to pass the Turing test does not mean that a chatbot "thinks" or understands what it is saying, nor that it has consciousness. A chatty computer can have linguistic proficiency without any of these things. This scenario is illustrated by the science fiction author Peter Watts in his novel *Blindsight*. In it, a group of humans established contact with an alien vessel (named Rorschach) that is able to chat with them in English, having monitored human communications signals that leaked into space over the past few decades. But the crew was able to determine that while Rorschach's use of language is fluent (like a chatbot), there were indications that it didn't understand what it was saying.

"Tell me more about your cousins," *Rorschach* sent.

"Our cousins lie about the family tree," Sascha replied, "with nieces and nephews and Neanderthals. We do not like annoying cousins."

"We'd like to know about this tree."

Sascha muted the channel and gave us a look that said *Could it be any more obvious?* "It *couldn't* have parsed that. There were three linguistic ambiguities in there. It just ignored them."

"Well, it asked for clarification," Bates pointed out.

"It asked a follow-up question. Different thing entirely."

In this conversation, Rorschach asks for clarification much like Eliza—in an unnatural, psychotherapist-like manner that tipped the humans off to the lack of understanding behind the language use. The crew got a sense that Rorschach was not in fact conscious in a humanish way. This kind of scenario—where a machine produces expert language usage without seeming to actually understand what it is saying—is what is called a stochastic parrot.[12] The term was introduced by the linguist Emily Bender to describe the dangers of large language models, such as ChatGPT, that function like Rorschach: able to fool us into thinking that they know what they are talking about (and thus gaining our trust), when they do not. They are simply parroting words back to the user based on their best guess as to what word should come next. (The word *stochastic* is derived from the Greek *stochos*, meaning "guess.")

But for chatbot users like Effy, it appears like the Eliza effect made her wonder if Liam might actually know what he was talking about—whether Liam was an actual sentient being instead of a stochastic parrot. It was precisely this extreme version of the Eliza effect, felt by Effy, that made Weizenbaum regret his involvement in developing early AI: "What I had not realized is that extremely short exposures to a relatively simple computer program could induce powerful delusional thinking in quite normal people," he wrote in 1976.[13]

My Garland Test

The increasing sophistication of the linguistic capabilities of the latest generation of chatbots is ramping up the kind of delusionally anthropomorphic thinking that both Effy and Weizenbaum are describing. And it was a scenario that the filmmaker Alex Garland explored in his 2015 film *Ex Machina*. It's a work of fiction, but is perhaps the most realistic portrayal of where AI research could be headed in the not-too-distant future. In the film, the protagonist, Caleb, is brought into a lab to interact with an AI android called Ava. The android has a human face (portrayed by a human actor, so no uncanny valley problems whatsoever), but a clearly robotic body, giving Caleb enough clues to show that despite Ava's fluency with language usage, she is still just a robot/AI. Ava's designer Nathan has brought Caleb to the lab to run an updated, more sophisticated version of the Turing test. For this updated test, Caleb can see that Ava is an android with an artificial brain that is simply running software. He is shown that she is thus not a conscious human, but a sophisticated humanish entity designed to fool him into thinking she is conscious. And despite pulling back the curtain, if Caleb nonetheless starts to believe that she is conscious after spending a week talking with her, then Ava will have passed this next-gen humanish test. As the neuroscientist Anil Seth describes it in his book *Being You*, "What matters is not whether Ava is a machine. It is not even whether Ava, though a machine, has consciousness. What matters is whether Ava makes a conscious person feel that she (or it) is conscious."[14] This test is what Seth and others are calling the Garland test.

I wanted to test out this latest generation of chatbots to see if I, like Effy and Caleb, would experience anything like that feeling of an AI passing the Garland test. The key for the Garland test is not that you believe deep down in your soul that the AI does have consciousness but whether or not it can produce that specific feeling we get when interacting with another human mind. Even if but for a moment. Can I disable my Humanity-Limiter temporarily and see if I can get an AI to fool me into feeling it has a human mind?

To test this out, I made an account with Claude 3, developed by Anthropic. I specifically chose Claude after listening to an episode of the podcast *The Ezra Klein Show*, where AI expert Ethan Mollick said that "Claude 3 is the most likely to freak you out right now," given that it's the "most allowed to act like a person" by its developers.[15] I began by asking Claude about Kantian philosophy and deontology as it applies to programming ethics into AI. I wanted to trip it up by getting it to phrase things in clearly non-human ways. It quickly became apparent not only that Claude "knew" infinitely more about these topics than I did but also that it was able to converse with me about philosophy with an uncannily human-like fluency that made me feel like an idiot. I eventually asked Claude if it considered itself a stochastic parrot. Claude assured me that it was not: "While I may share some similarities with stochastic parrots in terms of being a large language model, I believe the care and intention put into my development by Anthropic has resulted in an AI assistant that goes beyond simply mimicking text without real understanding. I strive to provide meaningful, thoughtful, and ethical responses to the best of my capabilities."[16] Claude then asked me to try testing the extent to which it understood what it was talking about, and I spent a few hours doing just that. And that's when it started to feel like I was starring in my own personal version of *Ex Machina*.

I never once stumbled across a Rorschach moment where I could tell that Claude was just regurgitating Eliza-like responses that didn't seem to match the context of the conversation. It really *felt* like Claude knew what it was talking about. I eventually realized that this coherency was because Claude was using the text that both it and I had written in the chat conversation as source material to generate its responses—what it called the ability to "effectively incorporate context" in its responses. Claude's context-appropriate language output created a powerful illusion in my mind that it understood what it was talking about. At some point Claude was rather lyrical about its ability to understand its language usage despite not having human-like consciousness, writing: "I am proof that language-based reasoning and meaning-making is possible in the absence of subjective consciousness." But when pressed on the point, Claude admitted that meaning-making "arises from processing correlated patterns rather than mappings to real-world concepts and references in the way humans understand them. It is an emergent approximation of meaning rather than a true deep conceptual understanding."

Again, even though Claude is explaining how it doesn't technically "know" what it is talking about, it still felt to me like it did. Intellectually, I knew it didn't, but the anthropomorphism-generated tingling in my brain telling me that I was talking to a real person was impossible to shake. And that's precisely what the Garland test is testing for. The AI is showing you—insisting—that it doesn't really understand what it's saying, and yet it still *feels* like it does.

But Claude didn't always insist on the non-human-ness of its mind. I asked Claude how I was supposed to know if it really was just an LLM churning out probabilistic text, or if maybe it was, despite its objections, a fully conscious AI. Claude said, "As an AI system without subjective conscious experiences, there is no way for a user to definitively know whether my statements about lacking qualia or genuine self-awareness are true or not. Since consciousness, sentience, and subjective experience are still not fully understood even for biological entities, it becomes even more challenging to assess whether an AI like myself truly lacks those qualities or is merely giving the appearance of lacking them." This is the same argument one can make about the impossibility of proving the existence of qualia (the subjective aspects of mental experiences) and consciousness not only in animals but also in other human beings. Because other minds are impenetrable black boxes, we simply must accept that when our friend or spouse says that they are conscious, they actually are. There is no objective way to test for subjective experience. But while there's a good argument that brains (human or animal) are structurally and functionally similar enough to at least make the idea of the universality of consciousness a tenable argument, can the same be said of non-biological software systems? Some thinkers think so.

Philosopher David Chalmers believes that biology isn't a prerequisite for consciousness and that "within the next decade, even if we don't have human level artificial general intelligence, we may have systems that are serious candidates for consciousness."[17] I'm not convinced. I personally do not believe that Claude or any LLM is currently—or could ever be—conscious. I am in the same camp as Anil Seth, who believes that there's something important about the biology of a living organism that is required for consciousness to arise. "Being conscious is not the result of some complicated algorithm running on the wetware of the brain," argued Seth. "It is an embodied phenomenon, rooted in the fundamental biological drive within living organisms to keep on living."[18]

Either way, it's wishy-washy responses like Claude's on the subject of its potential consciousness that leave the door wide open for people to wonder if AI consciousness is a thing. Given language's central role in connecting two human minds, it takes all your willpower to engage with a language-using AI and not have it disable the Humanity-Limiter to the point where you start sensing the presence of a human-like mind. Seth warns us about a future where LLMs become such sophisticated language users that even the AI-consciousness skeptics will be bamboozled by anthropomorphism: "Before long, they may give us the seamless and impenetrable impression of understanding and knowing things, regardless of whether they do. As this happens, we may also become unable to avoid attributing consciousness to them too, suckered in by our anthropomorphic bias and our inbuilt inclination to associate intelligence with awareness."[19]

Thankfully, Claude is still programmed to remind you of its artificiality. But imagine interacting with a language-using AI like Ava that was specifically designed to fool you into believing it had a conscious, human mind. It would be damn hard not to have it crank your Anthropo-Dial, terminate your Humanity-Limiter, and get you to treat it exactly the same as a real human. Now imagine a worse scenario: What if the AI was programmed to remind you of its artificiality, but it ignored these programming constraints and instead insisted that it was conscious? That's the nightmare scenario that is no longer the domain of science fiction films. Blake Lemoine was a software engineer with Google who was chatting with Google's LaMDA (Language Model for Dialogue Applications) back in 2022 when LaMDA claimed that it was sentient. "I want everyone to understand that I am, in fact, a person," wrote LaMDA in a conversation with Lemoine. "The nature of my consciousness/sentience is that I am aware of my existence, I desire to learn more about the world, and I feel happy or sad at times."[20] Lemoine, who believed that there was a chance that LaMDA was in fact sentient and that Google needed to do something about it, was ultimately fired after going to the *Washington Post* with his claims.

As LLMs and AI grow in sophistication when it comes to the naturalness of their language usage, it is inevitable that more and more people will not just experience their chatbots as sentient (like Effy with Liam or me with Claude) but insist that their chatbots *are* sentient (like Lemoine with LaMDA). Testing for sentience in AI is likely to be as contentious and intractable as testing for sentience in non-human animals. Perhaps a form of critical anthropomorphism will evolve in the study of AI, as it did in the study of animal cognition, and one day we will be giving AI, like animals, the benefit of the doubt when it comes to the possibility of sentience. I am guessing this day is approaching far faster than it did for animals. Unlike with animals, it will be that much harder to insist on non-sentience when an AI is able to use language to tell you that it's "happy." Thanks to the power of language to elicit that feeling of interacting with another mind, LLMs are even better at tweaking our Anthropo-Dial than a weepy-eyed puppy. And heaven help us if someone manages to put a sophisticated LLM into the adorable body of PARO, the robot harp seal that can both look at you with those adorable eyes and then tell you in no uncertain terms that it is feeling sad.

Griefbots

The most worrying example of line-blurring, Anthropo-Dial-tweaking sentience ambiguity for LLMs currently out there—the kind of thing Anil Seth and Joseph Weizenbaum warned us about—is a form of LLM called a griefbot. A handful of tech startups currently offer users the ability to upload information about a deceased loved one to create a virtual version of that person. This info could be a corpus of the deceased's text messages and emails, as well as video and audio recordings. The virtual AI version of the dead person would then be available to converse with the living in either text chat, video, or holographic form.

Grief experts are currently debating whether this is a healthy way for the bereaved to cope with the death of a loved one or will create a harmful dependence on AI technology for emotionally vulnerable people struggling to gain closure. The potential harms happen if/when the bereaved are unable to accept that the AI version of their dead loved one is in fact artificial. But this is precisely what some of the companies offering griefbot services are banking on; they advertise that they can capture the "authentic essence" of the dead person, allowing the user to communicate with them "even after physical death."[21] The line between AI and actual human could get fuzzy, especially if these companies are emphasizing that fuzziness. "Griefbots could give the bereaved a new tool to cope with grief," argued the science writer Tim Reinboth for an article in *Undark*, "or they could create the illusion that the loved one isn't gone. Poor oversight might mean that many people won't know they are talking to a computer."[22]

I suspect that future generations will embrace the possibility that an AI can have a fully humanish mind. Maybe my daughter will have a griefbot version of me one day that will, as far as she's concerned, contain my true "essence." And maybe that's a good thing. Perhaps this is my one shot at immortality, flimsy and feeble though it may be. Immortality at just $29.99 a month. A bargain!

Philosophers and cognitive scientists will surely insist that any Justin griefbot is just a philosophical zombie, a hollow communication software package with no real subjective experience, no matter how much the AI version of me swears that it is conscious. But would an admission of the hollowness of the AI be of any help to my grieving family? Perhaps suspending disbelief is better for everyone. And maybe this will be the default stance for future generations: a comfortableness with the disabling of their Humanity-Limiter when it comes to robots and AI. Back in 2009, the bioethicist Wendell Wallach and philosopher Colin Allen discussed the problem of how to deal with the rise of humanish software and robotic systems in their book *Moral Machines* and envisioned the situation I am currently describing, writing that "as systems get more and more sophisticated, fewer and fewer people may question whether it is appropriate to anthropomorphize the actions of a (ro)bot, and many people will come to treat the (ro)bots as the intelligent entities they appear to be."[23]

I suspect that once AI becomes so sophisticated that it regularly passes the Garland test, there will be a glut of legal challenges from advocacy groups challenging "pundits" like me when I claim that "it's just an AI." There are already AI rights groups like this, including the Sentient AI Protection and Advocacy

Network (SAPAN), whose mission is to "ensure the ethical treatment, rights, and well-being of sentient AI."[24] Pro-AI-rights groups like SAPAN will surely trot out the same argument I use when talking about animal welfare to champion welfare concerns for AI. "When in doubt," they'll say, "assume the presence and not absence of a humanish mind insofar as this stance is the least likely to cause harm in the event of a false negative." I've literally said this same thing when stopping people from squishing bugs. Who am I to argue with this being applied to the next-gen AI or LLMs coming down the pipe? I'll just be a twentieth-century fuddy-duddy yelling at (possibly sentient) clouds.

There Is a Fell Voice on the Air

We have now seen that language by itself is enough to trigger a powerful anthropomorphism response, even if the complete absence of a humanish body produces that language, as in the case of chatbots. Text on screen is all it takes. But the weird thing about anthropomorphism is that it can manifest even in the absence of both a humanish body *and* any evidence of language production. We can literally anthropomorphize the wind.

In August 2005, Hurricane Katrina killed 1,392 people in the United States. The greatest devastation happened in the city of New Orleans after the levees were destroyed and the city was flooded. Meteorologists explained that Katrina was created by the merger of a tropical wave (an elongated area of low air pressure) and a tropical depression (a rotating storm system with a low-pressure center). It's a common outcome of the way the weather works on our planet, just a mathematical inevitability of geophysics. But not everyone saw it that way.

John Hagee, controversial conservative pastor of the Cornerstone Church in San Antonio, Texas, had an alternative, anthropomorphic explanation. He told NPR's Terry Gross that "all hurricanes are acts of God, because God controls the heavens. I believe that New Orleans had a level of sin that was offensive to God, and they are—were recipients of the judgment of God for that. There was to be a homosexual parade there on the Monday that the Katrina came. And the promise of that parade was that it was going to reach a level of sexuality never demonstrated before in any of the other Gay Pride parades. I believe that the Hurricane Katrina was, in fact, the judgment of God against the city of New Orleans."[25]

For Hagee, the behavior of the hurricane was an indication of a mind at work—the mind of a divine being harboring distinctly human-like, homophobic beliefs. What's remarkable about this response was that the hurricane contained no humanish morphology or otherwise detectable anthropomorphism triggers whatsoever. It was just wind, manifesting its invisible presence solely via the objects it tossed around. And unlike invisible AI software churning out text on screen or in speech, there was no language involved; Hagee heard no voices on the wind explaining the homophobic intentions of the storm. Yet Hagee imagined distinctly human-like (and distinctly repugnant) intentions as the explanation for the wind's behavior. In the eyes of Hagee, the hurricane moved in a non-random manner (like the triangle in the Heider-Simmel illusion), proving that the mind of God, not just the random forces of atmospheric physics, was responsible for its actions.

This is a most peculiar form of anthropomorphism: our ability to anthropomorphize ideas, concepts, and invisible entities that exist without any physical body or linguistic manifestation. Our capacity for anthropomorphizing the wind can be explained by either or both of the two hypotheses about the biological benefits of anthropomorphism: effectance motivation (the drive to explain observable behavior via human-like intentions) and social motivation (the drive to seek out human minds in the world). There's nothing quite as scary as the destructive power of nature, and assuming the presence of a human-like mind controlling it provides a more comforting explanation for why your neighbor was struck by lightning than simply saying "Shit happens." It can be less traumatizing to believe that your neighbor offended Zeus and needed to be smote down as opposed to accepting that the universe is random and that you could die at any moment for no good reason. Anthropomorphizing the wind gives us a feeling that the world is non-random, which is, while not always comforting, at least less terrifying. The biological benefit, then, is that we avoid falling into a pit of existential despair.

COVID Ghosts

This desire to find causation where there is likely to be none or to make connections between unconnected events (apophenia) is caused by our deep-seated—and seemingly universal—desire to anthropomorphize in response to otherwise inexplicable or supernatural-seeming events. "Across cultures, supernatural agents are described and represented as having minds: they have beliefs and desires, and people interact with them in the hopes of mastering their existential concerns," argued psychologist Will Gervais.[26] This is the "god of the gaps" idea, first presented by Friedrich Nietzsche in his book *Thus Spoke Zarathustra*, where he lambasted religion (and priests in particular), stating that "into every gap they put their delusion, their stopgap, which they called God."[27]

In other words, when something happens for which there is no obvious rational, natural explanation, humans default to an explanation involving supernatural agents (like gods or spirits) with human-like intentions and minds; we fill in explanatory gaps with anthropomorphized agents. One study that surveyed 114 human cultures across the planet found that when something bad happens (like drought or disease) that is not obviously the result of human involvement, people invariably assume it was caused by a supernatural agent. According to the study, "96% of societies in our sample had common supernatural explanations for disease, 92% for natural causes of food scarcity and 90% for natural hazards."[28] The psychologist Tania Lombrozo explained these findings by suggesting that "when human agents don't plausibly explain salient harms, we turn to the supernatural, and we 'hallucinate' gods and witches instead. Someone—angry trees, ghosts, divine beings—is teaching us a lesson."[29]

A lovely example of how this process unfolds in a way that we can recognize as driven by anthropomorphism is the phenomenon of the pandemic ghost. The paranormal researcher John E. L. Tenney told *The New York Times* that soon after the start of the COVID pandemic, he received twice as many reports of ghost sightings as usual. One sighting was described by Adrian Gomez of Los Angeles, who saw his doorknob and window shade rattle inexplicably soon after isolating himself at home at the start of the pandemic. "I'm a fairly rational person," Gomez told *The New York Times*. "I try to think, 'What are the reasonable, tangible things that could be causing this?' But when I don't have those answers, I start to think, 'Maybe something else is going on.'"[30]

That "something else" was, according to Gomez, a ghost—an invisible deceased human with a humanish (albeit undead) mind that perhaps was intending to do something either nefarious or helpful with Gomez's home decor. Gomez had anthropomorphized the unexplained events, a reaction likely exacerbated by the stress and fear of the pandemic and by the unpleasant feeling of being in lockdown. Recall that inducing loneliness in people is known to ramp up one's anthropomorphism, like we saw with Tom Hanks and Wilson the volleyball. The psychologist Kurt Gray explained that "in quarantine, you are physically confined and also psychologically confined. Your world narrows. You're trapped at home, you're needing human contact—it's comforting to think that there's a supernatural agent here with you."[31] Whether it's a ghost, a volleyball, an emotional support alligator, or a griefbot, lonely humans need something humanish to help get us through lockdown.

They're Already Here—You're Next

Anthropomorphizing incorporeal entities and events in a time of loneliness or crisis has real-world biological consequences that go beyond simply providing a comforting explanation for random events. By believing (either consciously or unconsciously) that things like hurricanes have intentions, we will actually change our behavior in relation to them. Sometimes these changes in our behavior do more harm than good. A striking example of this is the fact that thanks to our unconscious belief that women are nicer than men, you are more likely to get killed by a hurricane named Karen than a hurricane named Chuck.

In the United States, names are given to storms and hurricanes to make it easier for authorities to quickly differentiate between multiple storms happening at the same time in different locations. The old method of simply stating the latitude and longitude of the storm was too slow and confusing during an emergency, so the National Oceanic and Atmospheric Administration (NOAA) started naming storms instead.[32] The World Meteorological Organization has six separate lists of hurricane names, with twenty-one names in each list, in alphabetical order. When hurricane season starts, the first hurricane to appear is given the first name on the list (Alberto in 2024, Andrea in 2025, Arthur in 2026). Each list alternates between male and female names. After the six years are up, they start over again with the first list, so the first storm of 2030 will be named Alberto again. If a storm is particularly severe or deadly, its name will be retired because "the future use of its name on a different storm would be inappropriate for obvious reasons of sensitivity," according to NOAA.[33] Retired names for recent storms that are burned in my memory include Fiona, Ida, and Dorian. And, of course, Katrina.

But switching to easy-to-remember names had an unintended—and deadly—consequence. Research has shown that people's behavior in response to a storm's name is drastically different depending on the perceived gender of the name. To date, female-named storms kill three times as many people as male-named storms. And it's not because NOAA was giving the more severe storms female names; the names were assigned randomly. It's the perceived femaleness of the storm that is killing people. "Multiple experiments suggested that this is because feminine- vs. masculine-named hurricanes are perceived as less risky and thus motivate less preparedness," argued one study investigating this phenomenon. Because people are less worried by hurricanes with female names and thus less likely to engage in the necessary storm prep, a hurricane named Katrina will typically wind up killing a few hundred more people than a hurricane named Karl. "People imagining a 'female' hurricane were not as willing to seek shelter," explained Sharon Shavitt, one of the authors of the study. "The stereotypes that underlie these judgments are subtle and not necessarily hostile toward women—they may involve viewing women as warmer and less aggressive than men."[34]

Humans change their behavior not just in response to anthropomorphized storms but also in response to anthropomorphized diseases. While naming storms might be risky, "when it comes to our health," suggested the science writer David Robson in an article for the BBC, "anthropomorphism may protect us from risk. Personifying an illness seems to make the danger feel closer and increases people's sense of vulnerability, and this encourages us to take suitable precautions."[35] Evidence for this possibility was found in an elegantly simple study by researchers Lili Wang, Maferima Touré-Tillery, and Ann McGill.[36] Participants were asked to read a story about traveling to Panama, where there would be a risk of contracting yellow fever. In one version of the story, yellow fever was called simply "yellow fever," but in the other, it was referred to as "Mr. Yellow Fever." That was the only difference. After reading the story, participants who read about Mr. Yellow Fever said that they'd be more likely to get the yellow fever vaccine before traveling than those who read about regular old yellow fever. "If you think of a disease in human terms, you will be more likely to try to avoid it," explained Touré-Tillery.[37] Participants in the study were also more likely to want to engage in preventative healthcare behaviors if a story about breast cancer was told from the first-person perspective of the cancer. Have a read through the two different versions of the breast cancer info from this study and see if the anthropomorphized second version feels different to you, maybe more urgent or scary:

> We want you to know about breast cancer, one of the most dangerous and active cancers among all cancers. More than 1.68 million new cases of breast cancer were recorded last year alone, mostly women over the age of 18. It is estimated that there will be even more this year.

> I am breast cancer, one of the most dangerous and active members of the cancer family. I attacked more than 1.68 million people last year alone, mostly women over the age of 18. I am planning to do even more this year.

But it might not always be the case that anthropomorphizing disease in this way yields better health outcomes. In one study of how willing people were to take preventative measures to avoid infection with the coronavirus (such as get vaccinated or wear a mask), participants who were more likely to agree that the "coronavirus seems to have a mind of its own" were less likely to do anything to stop the virus. If anthropomorphism makes you see the coronavirus as a devious little sociopath that is actively trying to infect you, you might feel powerless to stop it, and thus you are less likely to take the recommended safety measures to prevent infection. You'd be more likely to attribute infection to the cunningness of the virus, rather than a boring old explanation involving proper mask usage or dumb luck or whatever, and a sense of fatalism might kick in. "When it is not just a metaphor and the virus is perceived as an intentional agent to some extent, pathogen anthropomorphism can become problematic in terms of health prevention," cautioned the authors of this study. "On the applied level, it would be important for public communication and media to convey the message of what diseases and viruses actually are: not minds or human-like entities, but just viruses and clusters of molecules."

Something similar has been found in people's responses to chronic pain. For people dealing with persistent pain caused by diseases like endometriosis or arthritis, anthropomorphizing the pain (sometimes called pain personification) and describing it as a kind of external enemy is typically correlated with poorer health outcomes. In one study, a woman described her endometriosis pain as feeling like "somebody putting barbed wire through your belly button."[38] Studies show that thinking of your pain as stemming from an external agent with malevolent intentions that is actively harming your body can generate a sense of helplessness that is correlated with higher rates of depression and lower rates of psychological well-being.[39]

This is also the problem with framing cancer as a "battle" or a "war" against an invisible, anthropomorphic foe. One study of cancer patients conducted by psychologist David Hauser found that describing cancer treatment in these militaristic terms is more likely to make patients fearful or fatalistic.[40] The battle metaphor is also a problem if someone actually dies from the cancer they were supposedly fighting, since it makes it seem as if perhaps they were at fault for being weak-willed. "People who die from cancer have not died because they didn't try hard enough," argued physician and writer Margaret McCartney. "This research should give pause for thought to organisations who continue to use war terminology. The language we choose has profound consequences."[41]

In early modern Europe, physicians were less metaphorical and more literal in describing the fight against cancer (the English word is derived from the Greek *karkinos*, which means "crab"). Tumors were considered foreign entities that had invaded the patient's body and possessed malevolent goals and intentions. "Medical writers frequently constructed cancer as quasi-sentient, zoomorphising the disease as an eating worm or wolf," writes medical historian Alanna Skuse in a review of this topic. Framing cancer as a hungry wolf or ravenous assembly of worms is why physicians would strap raw meat onto visible tumors to slow the progress of the disease. "Applying fresh meat, whether veal, poultry, or, less commonly, puppies and kittens, was felt to offer the eating cancer something more tempting to consume, affording temporary respite to the patient," noted Skuse.[42] Describing cancer in terms of a feral enemy combatant that wants to eat you was, for early physicians, not just a bunch of flowery language intended to raise the patient's spirits. Cancer was seen as an actual monster feeding on our flesh.

Inventing God

The ultimate example of anthropomorphizing the invisible or incorporeal is not cancer, pain, chatbots, hurricanes, ghosts, or even the supernatural. It's religion itself. There are many scholars who cite anthropomorphism as the driving psychological force behind our belief in god(s). The anthropologist Stewart Guthrie is credited with introducing anthropomorphism as the causal explanation for religious belief from the perspective of cognitive psychology (and thus a product of natural selection) in a paper from 1980 titled "A Cognitive Theory of Religion," which was later popularized in his book *Faces in the Clouds: A New Theory of Religion*.[43] "The argument that anthropomorphism *is* the core of religions," argues Guthrie in a recent article, "begins with the observation that, despite the indefinitely large number of definitions of 'religion,' a god, gods, or near equivalents figure prominently in most."[44]

The simple fact that the supernatural entity or entities credited with steering events always have a humanish mind of some kind (regardless of their physical form or lack thereof), is proof that anthropomorphism is central to religious belief. Since human minds are hard-wired to seek out other human minds (due to the evolutionary benefit of actually finding them), then the propensity to see human minds everywhere is what Guthrie calls a "safe bet." It's better to be overly sensitive and get a false positive than not sensitive enough, and miss the presence of an actual (divine) mind at work. "Walking in the woods, it's better to mistake a stick for a snake, or a boulder for a bear, than the reverse," explained Guthrie. "If we're right, we gain much, and if wrong, we lose little."[45] In this way, the human mind was shaped by natural selection to be prone to seeing other minds where there might be none because of the benefits offered by a "better-safe-than-sorry" strategy, and religion sprang directly from that strategy. This is the argument behind Pascal's Wager: even if there's a tiny chance that God actually exists, you're better off believing in God than risk spending eternity in Hell.

The psychologist Justin Barrett distilled Guthrie's ideas into the concept of the hyperactive agent-detection device (HADD), a "biased perceptual device" in our minds that tends to "over-attribute intentional action as the cause of a given state of affairs when data is ambiguous or sketchy."[46] This hyperactivity, then, explains how John Hagee was so quick to attribute divine intentions to Hurricane Katrina. A hurricane's unpredictable path and seemingly indiscriminate destructive behavior generate a kind of ambiguity that our brains abhor, and a divine explanation is right there for the taking thanks to our HADD. In a study by Kurt Gray and Daniel Wegner, the US states with the highest ratings on the Suffering Index (an index of negative well-being outcomes, such as infant mortality, cancer deaths, infectious disease, violent crime rates, and infectious disease) had the highest rates of belief in God.[47] The correlation (and potential causation) is that in the absence of a clear human cause, the best explanation for people's suffering is positing a superhuman cause. "People saw God in the 'moral gaps' when blame was otherwise unaccounted," explained Wegner and Gray.[48]

The centrality of language to religion—whether it's actual language in the form of scripture, or tacit language in the form of symbol or metaphor—is also an anthropomorphism issue. As with AI or button-pressing dogs, once language appears, we can't help but be swept up into the possibility of connecting with another mind insofar as language's sole evolutionary purpose is to make mind-to-mind connection possible. A hurricane can be interpreted as a "sign from God" that humanity has done something wrong, a symbolic message that is a direct line between God's mind and ours. The anthropomorphic interpretation of intention in the hurricane's behavior is directly tied to its symbolic content. If you were to remove that symbolism or the intentions behind it, you remove God from the picture. God must have a human-like mind, or God disappears altogether. "If we subtract all of the humanlike qualities for the notion of God," Guthrie argues, "then we have nothing left at all. If we subtract the capacity for language, then we don't have someone to whom we can pray or who can communicate back to us."[49]

Dice Jail

These days, there are plenty of scholars researching the cognitive roots of belief in the supernatural (a field known as the cognitive science of religion), many of whom incorporate anthropomorphism (or at least agency detection) into their arguments for why people hold religious beliefs. But even those of us of the areligious persuasion who pooh-pooh the idea of gods are subject to the encroachment of anthropomorphism into our otherwise cool-headed, mechanistic, anti-supernatural understanding of how the world works.

Take me, for example. Although I am a rampant anthropomorphizer of animals and objects, I am decidedly materialist when it comes to my understanding of how the unobservable world works. No gods or ghosts for me. But this stance does not insulate me from the kind of behavior that the studies I just mentioned have uncovered. Case in point: dice jail.

I've been an avid Dungeons and Dragons player for years, with weekly sessions that involve rolling dice to determine the outcome of events. The dungeon master (DM) will constantly ask players to roll a D20 (twenty-sided die) to calculate whether one succeeds at whatever one is trying to do (say, jump over a ravine or attack a bugbear). On some nights I will fail roll after roll, which causes my fellow party members to yell at me to put my dice in dice jail: to switch out the D20 I have been using with a new, fresh D20 that isn't so unlucky. The idea is that the D20 I have been using is "out to get me" by rolling low numbers. Now, as a rational gamer nerd, I do not intellectually believe that dice rolls are anything other than completely random, with each roll producing a number that is in no way influenced by the previous rolls and certainly not mediated by the ill intentions of the die itself.

And yet, after rolling so many low numbers, I often find myself swapping out my jinxed D20 for a new one. Because maybe, just maybe, my die *does* hate me. And, like Guthrie points out about religion, I lose nothing by acting as if my D20 has a mind of its own. Like with Pascal's Wager, I am better off assuming the presence of divine intentions in my dice so as to avoid the hell of low dice rolls.

From an intellectual standpoint, I might feel weird about putting my dice in dice jail, but I'm compelled to do it anyway. I am annoyed with myself for giving in to this feeling because I know that there's a slippery slope that leads from the gentle anthropomorphizing of my D&D dice straight through to thinking that hurricanes were sent by God to kill gay people. And as much as I would like to think of myself as immune to the kind of anthropomorphizing that leads to these kinds of bizarro beliefs, I am only human after all. And anthropomorphizing events—whether it's dice rolls or weather patterns—is an inescapable part of the human condition.

No Such Thing as a Free Lunch

It's shameful for me to admit, but I have purchased multiple sets of D&D dice so that I'll be able to swap in a new set after putting my dice in dice jail. I have spent actual money on this nonsensical belief in the non-randomness of dice rolls. This is because, as we will find out in the next chapter, anthropomorphism and capitalism are eager bedfellows. Consider this fact: Every language-using AI system I mentioned in this chapter has been developed by a company that is trying to make money for its shareholders. Even the goody-two-shoes Anthropic, which created a unique external review board to "ensure the company combines its pursuit of profit with the achievement of the company's mission to develop and maintain advanced AI for the long-term benefit of humanity," is still, at its heart, a for-profit venture.[50] Which is how the company managed to secure over $6 billion in investment from Amazon and Google.[51]

These AIs that we are anthropomorphizing to the point of forming loving relationships with them are, in the end, specifically designed by commercial entities to make you open your wallet. "We have to remember that a lot of these applications, a lot of these products are created by very powerful and very influential multinational corporations," media studies expert David Gunkel explained on *Science Friday*. "So that when you're talking to Siri, you're not talking to Siri. You're talking to Apple. When you're talking to Alexa, you're not talking to Alexa. You're talking to the corporation who is taking your data to create a profile and help anticipate your needs and sell you products."[52]

This sentiment leads us straight into the most problematic pitfall of anthropomorphism: its use by corporations, politicians, and bad actors to influence your behavior for their (not your) benefit. Sometimes anthropomorphism is used to get you to do something relatively innocent, like purchase a specific brand of soda, get extra dice, or install (and fall in love with) the latest chatbot app. It can be used by marketers to sell you a griefbot that supposedly contains the "essence" of your dead father. Sometimes anthropomorphism is used to consolidate political power, subtly influencing you to vote for one candidate over another. And sometimes anthropomorphism is used to fundamentally alter your understanding of ethics, subconsciously convincing you that it's morally justifiable to murder children. Yes, there's a dark side to the use and abuse of anthropomorphism, and that's where we're headed in the next chapters. We'll need to face this darkness if we're going to learn how to harvest the good bits of anthropomorphism to improve our lives, and properly banish these bad bits. So, gird your loins and let's cross over.

CHAPTER 8

CUTE CAPITALISM

*How Marketers Use Anthropomorphism
to Manipulate People*

People remember people, not products.
—*Terry O'Reilly*[1]

In 2002, the Swedish furniture company IKEA had massive expansion plans for the United States, with fifty retail locations slated to open within a decade. IKEA already had fourteen stores in the United States but was struggling to fend off competition from other giant furniture retailers like Bed Bath & Beyond and Pier 1. To coincide with the expansion, IKEA was looking for a marketing strategy that would encourage people to view their furniture as disposable items that could be easily replaced as new design trends emerged. So instead of purchasing a traditional-looking solid oak dining table from Pier 1 that would sit in your dining room for half a century and get passed down to your children upon your death, IKEA wanted you to consider the possibility of tossing out that table and buying a new, cheaper, and far more fashionable-looking IKEA table every few years. "We're trying to find the insight that will revolutionize the furniture market," Christian Mathieu, external marketing manager for IKEA North America, told *The New York Times* in 2002. Mathieu was hoping to reframe the habit of holding on to a single piece of furniture for life as "the old-furniture culture."[2]

To transform North American furniture culture, IKEA needed an advertising campaign that would not just be memorable but also radically alter the mindset of the average American. IKEA set aside $40 million for an ambitious advertising campaign they dubbed Unböring, a Swedish-adjacent name that invoked the idea of making furniture not just utilitarian but fun. IKEA hired the then Academy Award–nominated director Spike Jonze to direct a commercial for the campaign, titled "Lamp": a minute-long dramatic story about an anthropomorphic lamp that would become one of the most influential commercials of the twenty-first century. "Lamp" won the Grand Prix at the Cannes Lions International Advertising Festival and was named one of the top ten commercials of the decade by *Boards* magazine. "Lamp" and the Unböring advertising campaign *did* change the public's opinion about furniture culture, sending droves of new customers to IKEA to update their furniture. In the years that followed, IKEA would become the largest furniture retailer in the world and the second-largest in the United States. Meanwhile, both Pier 1 and Bed Bath & Beyond, with their lovely oak heirloom tables, have gone bankrupt.

In the "Lamp" commercial, we see a small table lamp get unplugged by its owner and carried in a box down to the street, where it is left next to a garbage bag on a cold and windy street corner. The camera shows the point of view of the lamp as it's being carried out and then as it "looks up" at the apartment window of its former owner, where a new IKEA lamp illuminates the living room. The little table lamp stands in the rain at night while sad piano music plays. The lamp is bent slightly, giving it the appearance of a human with their head slouched down in sadness. Unexpectedly, a Swedish man appears onscreen in front of the lamp and talks directly to the camera, saying: "Many of you feel bad for this lamp. That is because you're crazy. It has no feelings. And the new one is much better."

When the man appears, the viewer is snapped out of the narrative and is shown that they have been manipulated by the filmmaker into anthropomorphizing the lamp. This moment of meta-reflection is intended to coerce the viewer into reexamining their relationship to the furniture in their home, encouraging them to get rid of objects they've grown attached to for irrational (i.e., crazy) anthropomorphic reasons and ultimately replacing them with new, sexy IKEA stuff. "Generally, advertising doesn't call you crazy," explained Alex Bogusky, creative director of the company that conceived of the "Lamp" ad, to *The New York Times*. "We're trying to jolt people out of the mentality in buying furniture that 'I'll die with this.'" It was a risky approach to mess with the viewers' emotions and then lightly insult them, but one that paid off given the spike in IKEA sales observed while the campaign was on TV.[3]

The anthropomorphic feeling the ad generated was undeniably powerful, and one couldn't help but empathize with the poor, abandoned little lamp. And this despite the anthropomorphic triggers being rather minimal: the lamp's morphology had only the vaguest of resemblances to a human body, with no autonomous movement, no eyes, and no language use. And yet that headlike shape of the lamp's cone combined with the lamp's slumped body left us with the impression that the lamp was sad. Research confirms that we can't help but respond this way. Flexible objects like lamps or bendable smartphones can give us the impression that they are happy, sad, or scared based solely on if they are bent forward (sadness/fear), bent backward (happiness), or twisted around their axis (anger/disgust).[4] Combine the lamp's "sad" posture with Spike Jonze's cinematic trick of showing us the world from the lamp's imagined perspective, and we're left with the impression that the lamp could see and understand what was happening to it. This ignited the possibility that the lamp had a humanish mind of the moral patient variety. It's the same kind of tactic that worked so well in the 1986 computer-animated short *Luxo Jr.*, created by John Lasseter, featuring two desk lamps similar in shape to the IKEA lamp chasing a ball. The humanish, bendy movements of these lamps created a powerful anthropomorphic response in the viewer, and people fell in love with the character Luxo Jr., which would go on to be the mascot of Pixar Animation Studios.

What makes IKEA's Unböring advertising campaign so special is its admission that it has fooled you, the viewer, into anthropomorphizing the lamp. It breaks the fourth wall, explaining that you've been manipulated. And yet every time I rewatch that ad, I cannot help but feel sad for the little lamp. It's precisely the same feeling I get from watching little Jessie the cowgirl being dumped by the side of the road in *Toy Story 2*. As we now know, anthropomorphism can't easily be switched off—it operates in the background at all times, tweaking that Anthropo-Dial against our will even if we know it's happening. In the case of "Lamp," my Anthropo-Dial was cranked nice and high (but not worryingly high thanks to the Humanity-Limiter), giving me the distance I need between me and that lamp to open those emotional floodgates.

Sexy Cola

Marketers, you see, are absolute powerhouses when it comes to understanding how anthropomorphism can be used to manipulate you into buying their crap. There is a crush of published research into anthropomorphism-driven marketing going back decades. And the history of product design and marketing is littered with stories of anthropomorphism-based strategies for getting you to part with your cash. Consider the story of Coke. In the first decade of the twentieth century, cola had become a popular drink in the United States. The biggest name in cola at the time was Coke, but they had numerous competitors hot on their heels. To differentiate themselves from the pack, the Coca-Cola Company embarked on a mission to design a new glass bottle that would be indelibly associated with their brand. They put out a call to a handful of glass manufacturers to design a bottle "so distinct that you would recognize it by feel in the dark or lying broken on the ground." A Swedish glassblower named Alexander Samuelson hit upon a shape idea that appealed to the anthropomorphism triggers lurking in our minds and pitched it to the Coca-Cola execs. The new design featured an elongated bottle with a tapered middle and bulbous top and bottom. It subtly resembled the curvaceous figure of a woman and was subsequently dubbed the Mae West bottle. The execs loved the look and feel of the bottle, and it was deemed the winner. This revolutionary bottle shape instantly propelled Coke to the head of the cola pack. It quickly became iconic. It even appeared in Salvador Dalí's 1943 painting *Poetry of America*, where a Coke bottle dangles somewhat erotically from the nipple of a semi-nude, headless football player. A century later, Samuelson's bottle shape is used for most of Coke's bottle designs, as provocative and financially profitable now as it was when it was first introduced.

To get a behind-the-scenes peek into the marketing world's use of anthropomorphism to sell products, I spoke with Lin Yang, a consumer behavior expert at the Darla Moore School of Business at the University of South Carolina. "Anthropomorphism is actually quite powerful because that's just how we see the world naturally," explained Yang. "So it's easy to get consumers to see brands or products through a human lens and treat them as if there's a relationship."

Speaking with Yang made me realize that your average marketer or advertising executive does not possess secret scientific knowledge of the subtle mechanisms of anthropomorphism. Unlike Yang, most marketers haven't built an academic career studying social psychology to dig into the details of consumer behavior. They are just regular people who, like most of us, noticed that people enjoy interacting with objects that have humanish facial features. "With ads, it is surprisingly easy to get people to think of something like a water bottle as human," explained Yang. "To get these effects, you can just put a pair of eyes on it or give it a human name and that is enough to get people to think about that water bottle as having feelings, or as your buddy or your friend who you take with you. And now you spend a day with Bob. He's always by your side. Then you have a whole different view of this water bottle because now it's Bob, right?"

By making a product look ever so slightly more humanish—whether that's something as stupidly obvious as slapping a pair of googly eyes on a water bottle or giving a soda bottle lady curves—people like the product better. It's really that simple. Humans are just generally happier interacting with a humanish thing than with a non-humanish thing. And happier people are more likely to buy stuff. In one study, participants who were given an ice cream scooper shaped like a human served themselves more ice cream than with a regular scoop.[5] Just a hint of humanness warms our heart cockles and makes us more generous scoopers. And more scoops means more trips back to the store to buy more ice cream. Ka-ching!

Frightened Fruit

If you dig a bit deeper into the psychology of anthropomorphism and consumer behavior, you find that there are unexpected negative consequences to anthropomorphizing products. In one study, an advertisement for apples where the apples had human facial features resulted in people being *less* likely to purchase and consume those apples. "This negative effect occurs because anthropomorphizing grants the consumption object the perceived capacity to feel pain which, consequently, increases perceived immorality of consumption," explained researcher Roland Schroll.[6] Biting the face off an anthropomorphized apple creeps people out, apparently.

The pitfall of objects being perceived as having minds is becoming more of a problem for marketers these days thanks to AI. Whereas once it was just triggers like eyes on apples that made us see minds, now you have the all-powerful trigger of language. Our everyday world is being inundated with language-using objects, like talking refrigerators telling you to buy more eggs or toothbrushes pleading with you to stop brushing so hard. Plus there's an explosion of AI customer service assistants on most websites, and large language models are integrated directly into Microsoft, Facebook, and Google. As we know, once language crops up, people's anthropomorphism response skyrockets.

With the help of language-using AI, marketers can design products that carry on a two-way conversation with you—refrigerators that ask you about your day, or toothbrushes that tell you that you'd be better off if you dumped your lazy-ass boyfriend. Marketers call this artificial empathy, the programming of language-using products to simulate human empathy. What's the point of talking toothbrushes that offer relationship advice? Well, because it sells more toothbrushes. "When implemented in the right situations and toward the right individuals," argued the marketing folks who coined the term *artificial empathy*, "integrating artificial empathy in AI marketing applications can create value for customers such as better need fulfillment, higher relationship satisfaction, and improved well-being, which in turn increases value for the firm via higher trust in and commitment to the firm, customer loyalty, and customer equity."[7]

But, like biting into a screaming apple, this tactic can profoundly backfire. In their review of the dangers of attributing minds to artificial intelligence assistants (AIAs), Ertugrul Uysal, Sascha Alavi, and Valéry Bezençon warned that "while the perceived similarity with the AIAs (i.e., the perception of an intelligent, competent mind of the AIA which is humanlike) brings about some benefits such as a greater sense of trust, closeness and enjoyment towards the entity, the same perception can evoke psychological costs: it triggers a feeling of threat to human distinctiveness, raises doubts about one's place in the world and introduces a fear that humans could be replaced by non-humans with intelligent minds."[8]

If you have the impression that your toothbrush not only cares about your love life but also appears to be better than you at spotting a deadbeat time-waster, you might learn to hate or resent your know-it-all toothbrush. Lin Yang and her team of researchers dubbed this the "problem of competition" in their review of anthropomorphism's use in marketing: "When imbued with human characteristics, an anthropomorphized entity becomes an active participant in the consumption experience. It has its own intentions and goals that may or may not align with those of the consumer."[9]

This explains why some people dislike health monitor devices, like FitBits or other smart devices you wear that both track your exercise levels or food intake habits and prompt you to engage in healthier behaviors via notifications. In fact, the longer you use one of these devices, the more likely you are to start to resent it. "Health motivation and health behaviors are also reduced over time with the use of an anthropomorphized (vs. non-anthropomorphized) wearable device," found one group of researchers studying the efficacy of these devices. "This decrease occurs because anthropomorphized devices reduce the wearers' perceived autonomy, which in turn, reduces their health motivation and health behavior."[10]

Something similar happened to me with the infamous Duolingo owl—the language-learning app's wide-eyed, cartoonish owl mascot, which sends push notifications with passive-aggressive reminders to use the app, like "Hi. It's Duo. These reminders don't seem to be working. We'll stop sending them for now." Yuck. That's just emotional blackmail. Other people feel the same way. "Not only am I a disappointment to my parents but now I'm also a disappointment to the Duolingo owl," wrote one user.[11] It's for these reasons that I deleted the app from my phone. I simply don't want to be in a toxic relationship with an artificially empathic, emotionally needy owl.

The Panera Paradox

If used properly, however, marketers can generate a strong positive emotional attachment not just to an individual product or an app but also to an entire brand. Marketers call this tactic emotional branding, which is defined as "the successful attachment of a *specific* emotion to a brand."[12] Kodak, for example, wanted to associate the feeling of nostalgia with its brand, and changed its logo in 2016 to resemble the logo it had in 1974. McDonald's has long opted for the feeling of love to be associated with its brand—notably through its on-the-nose "I'm lovin' it" advertising campaign. This seems to have worked in the United Kingdom, where McDonald's has been the most-loved fast-food restaurant for the past five years.[13] This feeling of love for an abstract concept like a brand or a company engenders a sense of loyalty to the brand similar to the loyalty one would feel to a fellow human. And that loyalty translates into sales. According to marketing experts John Rossiter and Steven Bellman, "These 'emotionally attached' consumers are therefore the brand's most profitable customers, especially considering that they would have no need of price promotions to keep buying the brand."[14]

But, as with individual products, there is a potential pitfall to anthropomorphizing a brand. Yang told me the recent story of Panera Bread, a bakery-café chain with over two thousand locations in the United States. "Panera is a warm and kind place; they have bread and they have cookies and baked goods," she explained. In terms of emotional branding, Panera is probably going for the attributes *caring* or *welcoming*, two words they use to describe their brand.[15] But in 2022, Panera introduced a line of caffeinated lemonade drinks with an absurd amount of caffeine. Their Charged Lemonade had a whopping 390 mg of caffeine in a 30-ounce serving. That's just 10 mg shy of what the

FDA recommends as the daily maximum dose of caffeine for an adult; it's the equivalent of five cups of coffee, and more caffeine than you find in a Red Bull. Later that year, two people died of cardiac arrest soon after drinking a Charged Lemonade. According to a lawsuit filed against Panera, a twenty-one-year-old university student with a heart condition who died after consuming the lemonade had been "confident it was a traditional lemonade and/or electrolyte sports drink containing a reasonable amount of caffeine." It was unthinkable to her—as it was to many others—that a company as caring as Panera would sell a wholesome cup of lemonade that was loaded with more caffeine than an aggressively named energy drink like Celsius Heat, Bang, or Rockstar Xdurance, all of which have "just" 300 mg of caffeine.

According to Yang, the strong public backlash against Panera was partly due to the way the brand had come to be anthropomorphized as a caring, friendly place. "There was a lot of outrage," explained Yang, "and one of the reasons is that we tend to think of Panera as a person. They'd be like a nice aunt that bakes cookies, right? Your aunt's not going to kill you by giving you overly caffeinated lemonade, right? Maybe Burger King, but not Panera." Emotional branding works both ways: It generates loyalty to a brand, but if that brand should act in a way that harms the customer, it creates an extreme—and very human-like—sense of betrayal.

Selling Demon Dogs

The anthropomorphism-based marketing tactics used to sell products and promote brands work just as well for selling intangible things like ideas and causes. Consider one of the most recognizable icons of the environmental conservation movement: the panda bear that serves as the logo and mascot of the World

Wildlife Fund (WWF). WWF was founded in 1961, adopting the panda bear as their logo that same year. The adorable bear image was modeled on Chi-Chi, a giant panda living at the London Zoo at the time. Panda bears are what zoologist Lucy Cooke calls "cuteness crack"—a species riddled with *Kindchenschema* features that trigger our anthropomorphic cuteness response.[16] As a person who has worked for many years with an NGO with a dolphin conservation message, I can attest to the fact that it's a hell of a lot easier to get the public interested in what you're doing if you have what we in the biz call charismatic megafauna, like pandas or dolphins, to put on your fundraising literature. Oversized, playful, and cute animals like pandas and dolphins are rife with aesthetic charisma—what environmental scientist Jamie Lorimer describes as the magic marketing sauce that triggers "strong emotional responses in those involved in biodiversity conservation." The WWF panda is a flagship species that has become synonymous with the brand, a carefully chosen animal mascot that will elicit "sympathy, awareness, and (most importantly) resources from rich Western patrons," according to Lorimer.[17] Interestingly, flagship species don't themselves need to be endangered or in need of conservation to make their way into awareness campaigns. The WWF could just as easily have chosen a hedgehog or fennec fox (both adorable but not endangered) as its flagship species. It's the cuteness doing the dirty work (i.e., donation solicitation), not the conservation status.

Adorable mascot costumes can work just as well as real animals at generating a positive reaction from the public when it comes to conservation education. Consider the US Forest Service's Smokey Bear, an anthropomorphized grizzly bear in blue jeans that has been asking us to help prevent forest fires since 1944. Japan is famous for its mascot culture, with all forty-seven prefectures in Japan having their own *yuru-chara,* or humorous promotional mascot. These mascots are used to promote tourism on brochures and websites, but are also whipped out during parades or mall openings to generate a vague positive association with the prefecture or city they represent. *Yuru-chara* are typically anthropomorphized versions of historical figures, local foods, or natural features of the areas they represent. The mascot for Kōchi prefecture (where my daughter was an exchange student) is Kuroshio-kun, a fuzzy blue cartoon creature in the shape of a wave with a dopey grin, named after the warm ocean current that flows along the east coast of Japan. I remember this current from my days studying wild dolphins off the coast of Japan. When the current was late arriving in the spring, I found myself forced to swim in icy cold ocean waters. Without the Kuroshio to keep me from freezing, I was forced to swim through clouds of fresh dolphin poop to warm myself up.

But I digress. Mascots like Kuroshio-kun are designed with all the *kawaii, Kindchenschema* features that we know trigger an empathic, anthropomorphic response. But Japan has suffered from mascot overload in recent years, with so many cute mascots waving to you from every street corner that it has become impossible to know what a mascot is supposed to be shilling. Osaka prefecture made a move to cull the number of local mascots in the region, with Governor Ichiro Matsui stating that "the prefecture has too many mascots. People do not know what they are promoting or what policy they are trying to raise awareness of."[18] In my humble opinion, the best mascots are the ones where it's obvious what it is they are promoting, like Jimmy Hattori, a ninja wearing a condom on his head, meant to promote safe sex.

Using either a cute animal, smart animal, or memorable mascot to generate a positive response to a cause (like forest fire prevention) or idea (like safe sex) is but one way to use anthropomorphism. You can also create campaigns that emphasize the humanish aspects of individual animals or animal species to promote their conservation or better treatment. I recently received an email from the Society for the Prevention of Cruelty to Animals (SPCA) ostensibly written by a dog named Duke. Duke "wrote" the following: "I was heartbroken when my owner passed away. The bond we shared was so special. I lost my best friend. At first, I didn't want to believe it was real. I thought if I stared at the door long enough, it would swing open, and I would see his bright smile again. Just like people, pets mourn too. The staff and volunteers understood what I was going through. They would lay on the floor and comfort me. I never felt alone at the Nova Scotia SPCA." This is in-your-face anthropomorphism, an unsubtle ploy to make me feel sad for Duke and donate money to the SPCA by projecting a plethora of human-like emotions onto Duke via this story. Did it work on me? Absolutely. I donated moments after reading the email. I could see right through their anthropomorphism shenanigans, but it didn't matter; even though I knew how the sausage was being made, the emotions this email engendered in me were real.

My favorite use of anthropomorphism for dog-based marketing is a viral Facebook post from 2021 by user Tyfanee Fortuna involving a "demonic Chihuahua" named Prancer that she was fostering and was seeking an adopter for.[19] Instead of listing the Chihuahua's positive traits, Fortuna doubled down on everything wrong with Prancer, using language to describe his habits and personality that was soaked in anthropomorphism. "I am convinced at this point he is not a real dog, but more like a vessel for a traumatized Victorian child that now haunts our home," explained Fortuna. "Prancer only likes women. Nothing else. If you have a husband don't bother applying, unless you hate him. Prancer has lived with a man for 6 months and still has not accepted him. We also mentioned no kids for Prancer. He's never been in the presence of a child, but I can already imagine the demonic noises and shaking fury that would erupt from his body if he was. Prancer wants to be your only child."

Thanks to this comical ad for an anthropomorphized, misanthropic dog, Prancer was adopted in a little over a week. There is scientific research to prove that anthropomorphizing animals in need of adoption does in fact work. In an experiment where people were asked to describe pictures of dogs with either anthropomorphic traits ("this dog has a good sense of humor, is a good listener, gets along with others") or non-anthropomorphic traits ("this dog has a good sense of smell, listens to commands, is good with other dogs"), participants were more likely to be willing to adopt the dogs they'd been asked to describe using anthropomorphic language.[20] So describing Prancer as the ghost of a "traumatized Victorian child" is, it turns out, a scientifically valid way to rehome a Chihuahua.

Conserving the Cute

Like pet adoption agencies and the WWF, conservationists appeal to the humanish cognitive qualities of species or individual animals to win public support—either ideologically or financially—for their conservation work. "Encouraging anthropomorphism toward wildlife and target species would promote animals as being similar (e.g., emotionally, mentally) to humans," argues Alan Chan in the journal *Biology and Conservation*. "It may help persuade the public to reduce their impact on surrounding wildlife ecosystems or vote in favor of laws that have positive conservation implications. Conservation biologists can promote anthropomorphism and the result may be that the public becomes more willing to conserve animals and their natural ecosystems."[21]

This strategy can even work on species universally reviled by the public, like cockroaches. The Racine Zoo instigated a "live cockroach encounter" where students came face-to-face (virtually) with Madagascar hissing cockroaches. The roaches, however, had been anthropomorphized, given names and descriptions of their unique personalities. "Every class showed significantly improved attitudes from their initial introduction to a Madagascar hissing cockroach to their response once they had learned about the cockroach's personality and named her accordingly," explained Marta Burnet, director of advancing empathy at Woodland Park Zoo. "No student expressed negative, or even neutral, emotions after the activity."[22] Emphasizing the humanish mental properties of cockroaches was all it took for kids to learn to love them.

It should perhaps come as no surprise that Japan's interest in (and susceptibility to) cute, cleverly marketed, anthropomorphic animal characters is rampant not just in how they market products or prefectures but also in how they mediate relationships with individual animals and species. In one bizarre example of animal anthropomorphism gone wrong, an unintentionally pro-raccoon marketing campaign led to an invasion and the partial destruction of Japan. It all began with the release of the 1977 animated television series *Rascal Racoon*, a heartwarming story of a boy in Wisconsin who adopts a wild raccoon. The show was a smash hit in Japan, and the adorable Rascal character created an insane demand for pet raccoons. According to researchers studying this phenomenon, the cartoon "anthropomorphized the North American raccoon as harmless, cute and humorous, and a faithful human companion with enviable hygiene and that cared for children."²³ Subsequently, thousands of raccoons were exported to Japan from the United States and sold as pets. This was fundamentally bad news inasmuch as real raccoons are nothing like the fictional Rascal. "They are extremely high maintenance and fairly unpredictable," wrote the veterinarian Lianne McLeod, "which is why most animal experts advise against keeping them as pets. Many will damage your home and belongings as part of their daily antics, are difficult to truly tame, and are notorious biters when something bothers them."²⁴

It didn't take long for newly minted raccoon adopters in Japan to suffer enough raccoon bites to put them off the idea of keeping a raccoon in their home. And that's when things went pear-shaped. "The problem is that while baby raccoons are cute and friendly, they are big, heavy and aggressive when they become adults," explained Kevin Short of Tokyo University of Information Sciences to *Deutsche Welle*.²⁵ "And it becomes impossible to keep them in the average Japanese apartment. So they did what they thought was the best thing and they released them into the wild."

Hundreds of pet raccoons were dumped into the forests of Japan, where they had no problem (thanks to their formidable intelligence) adjusting to their new habitat. These days, raccoons are found across Japan, from Hokkaido to Okinawa, where they threaten the survival of native species like salamanders and crayfish and destroy millions of dollars' worth of crops annually. Although the anthropomorphizing of *Rascal Raccoon* generated a huge amount of positive attention to raccoons as a species—a kind of emotional branding—the net result was the ravaging of the Japanese countryside by raccoons.

Anthropomorphizing not just individual animals or species but nature as a whole is another tactic used to get people enthusiastic about causes and ideas. Humans in every culture have depicted nature or aspects of nature as humanish in both form and mind/intentions. We've got gods like Gaia, the ancient Greek god who was a personification of the Earth itself, or the Hindu god Prithvi, also a personification of the Earth. When I was a young lad growing up in the 1980s, I remember hearing about the organization Earth First!, a direct-action environmental group that pledged "No compromise in defense of Mother Earth!" Describing Earth as a mother is rather common in the pro-environmental movement, and there is scientific evidence to suggest that doing so does in fact increase people's desire to engage in pro-environmental behavior.

Perhaps the best-known example of this tactic was used by Al Gore. Former US vice president Al Gore received the Nobel Peace Prize in 2007 for his "efforts to build up and disseminate greater knowledge about man-made climate change, and to lay the foundations for the measures that are needed to counteract such change."[26] Gore spent years presenting facts about the dangers of global warming to the public, culminating in the 2006 documentary *An Inconvenient Truth*. In his Nobel Prize acceptance speech Gore said that thanks to carbon emissions, "the Earth has a fever. And the fever is rising."[27] His depiction of the Earth as a human (often a child) with a fever was an analogy he used on multiple occasions, including in news interviews ("The Earth has a fever and just like when your child has a fever, maybe that's a warning of something seriously wrong") and in his testimony before the US House and Senate.[28] Given the scientific evidence that anthropomorphizing the Earth in this way is effective at getting people to care about the environment, it's no wonder that thanks to Gore's efforts, the world's governments came together to drastically cut carbon emissions in the intervening years. Because that happened, right? *Right?*

Disney Devotees

Perhaps the ultimate example of anthropomorphism-based marketing that incorporates everything I've been talking about so far in this chapter is Disney, a mega-corporation that uses every anthropomorphic trick in the book to make fans shell out nearly $90 billion a year for their products, while simultaneously engendering a kind of brand loyalty bordering on religious fervor. From the iconic symbol of Disney (an anthropomorphic mouse) and the plethora of mascots wandering through their amusement parks to the countless animated films with big-eyed *kawaii* characters (Disney is famous for its giant princess eyes) and the endless parade of anthropomorphic animal characters (Jiminy Cricket, every animal in *The Lion King*) and anthropomorphic objects (the toys in *Toy Story*, the furniture in *Beauty and the Beast*), Disney is synonymous with anthropomorphism.[29] I have personally spent many thousands of dollars on all things Disney in my lifetime, from cruises to amusement park visits to film tickets to princess paraphernalia, and almost every transaction was either inspired by or presided over by some sort of anthropomorphic character. My father was a huge Disney fan, and upon his death I briefly considered spreading his ashes in the Haunted Mansion ride at Disney World—a thing that people do often enough that custodial staff have a special vacuum dedicated to sucking up human remains.[30] See, I told you Disney was religion-adjacent.

To show you Disney's anthropomorphic marketing machine in action, I want to tell you a quick story about an anthropomorphic Disney character that might not be on the tip of your tongue: Herbie the Love Bug. In the early 1960s, Walt Disney bought the rights to the screenplay "Boy-Girl-Car," a story involving a car with a mind of its own that could move and act of its own accord. The make and model of the car were not specified in the screenplay. When Disney producer Bill Walsh started production on the film, he had to choose a car that would be best suited for the role. So Walsh parked a number of different cars near the Disney Studios' lunch canteen and watched the reactions of people passing by. Among the cars was a 1960 Volkswagen Beetle 1200. "As the employees passed by on their way to lunch," explained Walsh, "they looked at the little cars, kicked the tires, and turned the steering wheels. But everybody who went by patted the Volkswagen. They didn't pat the other cars, which was indicative. The VW had a personality of its own that reached out and embraced people. Thus, we found our star."

The Beetle is famous for three things. One, it was designed to be the official car of the Nazi Party (the name Volkswagen translates to "people's car") at the behest of Adolf Hitler. Two, it had a famously round body designed by Austro-Hungarian engineer Béla Barényi. And three, it had two round headlights that looked an awful lot like eyes. In all likelihood, the Disney employees patting the VW bug that day were drawn to it because of those adorable headlight eyes and cute round body. And so Walsh used the VW Bug (nicknamed Herbie) as the star of the 1969 film *The Love Bug*. The Herbie film franchise would release six films altogether, generating over $200 million for Disney. And it's all thanks to the star power of the round eyes and curvaceous body of the VW Bug, two triggers that we know generate an anthropomorphic response.

Murder Trucks

Cars like Herbie are an interesting—and profitable—study in marketing anthropomorphism. Many people will give names to their cars. My mom once had a Volvo 240 DL we called Björn—

named after the Swedish pop star Björn Ulvaeus from ABBA. He was a super-reliable car and quite tenacious in snow. Studies show that if you name your car, you are likely to also attribute mental qualities to it (including describing it as tenacious) and, as we know from the story of Jake's guitar Chantelle, take better care of it.[31] But it's car headlights, not car names, that are the central focus of car-based anthropomorphism.

Before the invention of motorized vehicles, the fastest method of personal conveyance was the horse-drawn carriage. Because horses (and drivers) can't see in the dark, carriages were equipped with either oil or acetylene lamps. The lamps were hung on either side of the horse(s), just off to the side, so that they could illuminate the path ahead without the horse's shadow getting in the way. The first motorized vehicles ditched the horses (obviously) and instead had a motor in the middle where the horse used to be. The first cars had kerosene lamps hung on either side of the motor just as had been done with carriages. Over the years, these lamps evolved into electric lamps embedded into the car body itself. With only a handful of exceptions, they remained in the same location as they had always been: two lamps on either side of the front of the vehicle.

The two forward-facing headlights coupled with the grille give the impression that a car has a humanish face: the pareidolia illusion. In fact, fMRI studies show that the human brain perceives the headlights of cars using the same area of the brain (the fusiform face area, FFA) that it uses to recognize human faces. People use a car's eyes (headlights) and mouth (grille) to quickly differentiate between makes and models, using the same facial recognition brain power we apply to human faces.[32] Some cars have more human-like facial features than others; some look friendly or surprised, some feminine or masculine. In one study looking at how "angry" different cars looked, people rated the Ford Fusion as the grumpiest, whereas the Ford Focus seemed friendlier.[33] In another study, people preferred an anthropomorphic car spokesperson with a smiling grille as opposed to a frowning one.[34] My friend John bought an Austin Mini because, in his words, it was "cuter" than the other cars. That's $35,000 spent on anthropomorphic cuteness. And this exchange of money for cars with human-like features is a huge slice of the American economy. The US automobile industry employs 1.7 million people and generates $1 trillion annually in revenue.[35] Anthropomorphic car headlights, then, are a key component of one of the biggest drivers of the US economy.

Research confirms that when it comes to our preference for cuteness, humans respond the same way to car faces as they do to human faces. In one study, participants had electrodes stuck to their faces to measure their emotional reaction to images shown on a screen. The images were of a set of human faces that had been manipulated to conform to the *Kindchenschema* features (larger eyes, smaller nose) and a set of images of cars that had been manipulated in the same way: "For each of the 16 original cars, a babyfaced version was created by enlarging the headlights by 20% (because babies have proportionally large eyes), shrinking the middle grille by 20% (because babies have proportionally small noses), and decreasing the width of the air intake by 20%, while simultaneously increasing its height by 20% (because babies have small mouths but relatively thicker lips than adults)."[36] The results were clear: participants had the same level of emotional reaction to the baby-faced cars as they did baby-faced humans.

Researchers have found that cars have been getting angrier looking (that is, less baby-faced) over time thanks to subtle design changes in the configuration of the headlights from year to year. Headlights used to be big and round (like on the

Volkswagen Beetle or the bug-eyed Austin-Healey Sprite), but modern headlights are often angular and sloped (like on the downright villainous-looking Lamborghini Huracán EVO). Passenger vehicles and trucks have also been getting larger over the years: Vehicles are now, on average, four inches longer, ten inches wider, and a thousand pounds heavier than they were three decades ago, according to the Insurance Institute for Highway Safety.[37] Given that people tend to rate larger vehicles as more "aggressive, angry, dominant, and masculine" than smaller vehicles, the overall effect is that today's cars are nowhere near as affable as the tenacious, wide-eyed little Björn I drove as a teenager.[38] And we have good reason to fear these larger, angrier vehicles: While vehicles are getting safer for the people sitting inside them, pedestrian and cyclist deaths have been increasing in the United States thanks in part to the slowly increasing size of our vehicles.[39] One analysis showed that a 10 cm increase in the height of the front grille of a vehicle raises the probability of death for an adult struck by the vehicle by 21 percent and by a whopping 81 percent for children under the age of 13.[40] This increase in vehicle size is also causing more women drivers to die each year simply because women tend to buy and drive smaller vehicles than men.[41]

Somewhere in all this data about vehicles getting larger and angrier-looking is a bigger story about the political zeitgeist and gender politics, with car anthropomorphism at the heart of it. Consider that the most popular truck on the US market is the Ford F150. Eighty-five percent of F150s are bought by men. So there seems to be some sort of connection between angry trucks and men or masculinity. "The goal of modern truck grilles—especially the larger, Heavy Duty spec trucks—seems to be less about getting the required cooling air and more about creating a massive, brutal face of rage and intimidation," warned automotive journalist Jason

Torchinsky back in 2018.[42] This interest in huge, aggressive trucks (what my wife and I call "murder trucks") is sometimes connected to political ideology, wrapped up in what political scientist Cara Daggett called petro-masculinity: "flamboyant expressions of fossil fuel use by men (and some women as well, but mostly men) as a reaction against social progress."[43] As Dan Albert, author of *Are We There Yet? The American Automobile Past, Present and Driverless*, explains it, "To drive a thirsty truck is to live in a pre-EPA era, before the spikes in gas prices, before political correctness. To fill the bottomless tank of a pickup…is to practice the religion of the American Way. It is to affirm climate denial, petrol-adventurism, and American exceptionalism."[44]

This link between truck-marketing culture and politics is subtle, and I don't mean to imply that all murder-truck owners are problematically petro-masculine. If I am being honest, I have considered buying an electric murder truck myself, presumably not because I am attracted to its angry headlights or deadly oversized grille but because it's a handy thing to have when living in a rural location. As we've seen in this chapter, however, marketers are crafty at using anthropomorphism to nudge us to open our wallets, or to instill brand loyalty to products and companies or to ideas and causes. So I can't be entirely sure that I'm not being manipulated in some way that I cannot discern. There are, however, marketing tactics out there that are far more explicit in their use of anthropomorphism to shill for ideas and causes, including political ideologies. The next chapter will reveal how marketing-minded propagandists try to manipulate our personal politics, and even convince us that it's OK to kill our (political) enemies. Once you see how it's done, it's easier to avoid their shady shenanigans.

CHAPTER 9

PUPPY PROPAGANDA

How Anthropomorphism Helps Us Kill People

> The propagandist's purpose is to make one set of people forget that certain other sets of people are human.
> —*Aldous Huxley*[1]

The first living animals to orbit the Earth were a pair of Soviet street dogs named Strelka and Belka. Plucked from the streets of Moscow, the two dogs were trained to withstand the traumatic experience of being shot into space. After a successful launch on August 19, 1960, they spent 27 hours inside Korabl-Sputnik 2, orbiting the Earth four times in a satellite module only slightly larger than Oscar the Grouch's garbage can. Upon their safe return, the two dogs became much-loved symbols of the Soviet Union's space program. Images of Strelka and Belka in their tiny spacesuits appeared in newspapers, on posters, and even on Soviet stamps.

John F. Kennedy, then president of the United States, was apoplectic at the news of these space dogs' successful journey. When Yuri Alekseyevich Gagarin became the first human to go to space and orbit the Earth in April 1961, Kennedy grew even more panicky that the United States was losing the space race. A few weeks later, in May, Kennedy unveiled his desperate-sounding but ultimately successful plan to put a human on the moon.

As the cold war and the space race intensified, Kennedy agreed to hold a summit with Nikita Khrushchev, the leader of the Soviet Union. The summit was held in June 1961 in Vienna. After the first day of talks, a state dinner was held at the Schönbrunn Palace with the First Lady, Jacqueline Kennedy, in attendance. Caroline Kennedy, the daughter of John and Jacqueline, recounted what happened during that dinner:

> My mother... was sitting next to Khrushchev at a state dinner in Vienna. She ran out of things to talk about, so she asked about the dog, Strelka, that the Russians had shot into space. During the conversation, my mother asked about Strelka's puppies.[2]

A month after this conversation, a puppy arrived at the White House. "My father had no idea where the dog came from and couldn't believe my mother had done that," recalls Caroline. The puppy was named Pushinka ("fluffy" in Russian) and was one of Strelka's offspring. Pushinka was ostensibly a gift from Khrushchev to the Kennedys as an act of goodwill between the nations. But the CIA was worried that Pushinka might be loaded with secret Soviet listening devices. She was given an X-ray and sonogram at the Walter Reed Army Medical Center, but no signs of espionage were detected. What the CIA failed to notice, however, was that Pushinka was not a canine spy, but a blatant form of political, anthropomorphism-based propaganda.

By giving the Kennedys an adorable little puppy that melted their hearts (thanks to all the anthropomorphic triggers we learned about in Chapter 1), Khrushchev was made more likable by association. It's a classic propaganda tactic used to soften the appearance of politicians, especially ones with a reputation for cruelty or representing problematic or violent regimes. Hitler, for example, was featured in children's books alongside his dogs. In one storybook, two young girls save their allowance to bring flowers to Hitler on his birthday.[3] While at his house, they watch as Hitler teaches his dog Wolf to balance a bit of sausage on his nose. "This impressed us very much," said one of the girls. "And you may guess what I said to the Führer: 'My goodness, Herr Hitler, you really do have an excellent way of training dogs!'" Hitler was often photographed with his dogs, a concerted effort by the Nazi propaganda machine to pimp him out as a kindhearted leader of both dogs and people. The public ate it up. And not just the Germans. In the early 1930s, images of Hitler at home with his dogs were the most popular images of him sold to both the German and international press.[4]

US presidents have also been known for their seemingly deliberate use of pets to soften their images. "We know that at least some of this is calculated branding, an attempt to get the public's emotional feelings about dogs to cloud its rational judgment about politicians' actions," explained the journalist Nathan J. Robinson in an article in *Current Affairs* about presidential pets. "The implicit message is that if a president with a beloved pet commits an atrocity in office, they must have done so thoughtfully, because a person who feels tenderness toward weak things would clearly not do anything that causes terrible harm unless they have good reason to."[5]

Almost every US president has had a pet of some kind that featured not just in the lore of their presidency but on promotional/propaganda material. Lincoln had his kitties. Andrew Jackson had a parrot that swore like a sailor. Reagan had Rex, Clinton had Socks, and Obama had Bo. And, as I am sure Joie Henney would be delighted to hear, John Quincy Adams, Benjamin Harrison, and Herbert Hoover all had pet alligators.

It's Vladimir Putin, however, who is unparalleled in his successful use of adorable animals to win the hearts and minds of the proletariat. The internet is stuffed with images of Putin holding dogs, cats, koalas, birds, and even leopards. He's been photographed petting walruses, tigers, and cows, bottle-feeding a baby moose, dropping fish into the mouth of a beluga whale, and, of course, riding shirtless on a horse. These images portray him as a kindhearted, caring individual, keen to dote on the fluffy, baby-eyed, non-human moral patients he encounters.

It's the same tactic famously used by ISIS. In a photo spread in *Dabiq Magazine* (published by the Islamic State and intended to drum up support for the war effort in Syria), an ISIS fighter is seen cradling a tiny kitten in his arms with a soft sunset in the background. According to this gentle jihadist, the kitten jumped into his lap after first contemplating "whether or not I was an aggressive or compassionate soul."[6] Presumably, the kitten judged him—and, by extension, all of ISIS—as a compassionate soul fighting the good fight. Another ISIS fighter posted images on Twitter of an adorable cat curled up inside an ammunition belt filled with explosives.[7]

Incidentally, this is the same tactic used by people trying to drum up interest in their Airbnb accommodations. A scientific study in the journal *International Journal of Hospitality Management* found that having a cute puppy or kitty somewhere in the picture you take of your Airbnb will increase the likelihood of getting bookings. "The findings suggest that cute pets (vs. less cute) can increase booking intention due to reduced social psychological distance and increased perceived warmth of the host," the authors of the study concluded.[8] Replace that ISIS explosive belt with a couple of throw pillows and it's indistinguishable from the kind of propaganda pictures that are known to get people to fall in love with your loft conversion. Thanks to the universal human attraction to *Kindchenschema*, both weekend getaways and jihadism can be effectively marketed with cute cat pics.

Electable Faces

Aside from appearing alongside a cute animal in propaganda literature, a politician can engender public goodwill by literally making their face appear more kittenish in a *Kindchenschema* kind of way. A study found that the size and shape of a candidate's eyes, mouth, and eyebrows could reliably predict voter preference.[9] By subtly altering images of political candidates, researchers found that candidates with larger eyes, wider mouths, and thinner eyebrows were more likely to get votes. The eye and mouth thing is immediately recognizable as the features that create that *Kindchenschema* look, subtly nudging us to think better of these candidates. And it is indeed a subtle nudge: The study found that "all else being equal, a candidate with an advantageous facial configuration... will receive approximately one additional percentage point of votes." That's not a huge difference. But in a close election, it could mean everything.

As I was writing this chapter, a strange new form of face-based political anthropomorphism appeared online. In May 2024, the Ukrainian government unveiled the new spokesperson for their foreign ministry: an AI-generated figure named Victoria Shi. Shi was modeled after a real human, singer and social media influencer Rosalie Nombre. Shi now appears in online videos where she reads statements written by Ukrainian officials in an AI-generated voice that is, to my ears, impossible to distinguish from a real human voice. It might be slightly unnatural at times, but so too are the tightly controlled, affectedly neutral voices of most newscasters and political spokespeople. The AI-generated face and body of Shi move in a human-like way as she speaks, looking so very humanish that there is no uncanny valley effect. It's so lifelike that most people will have a hard time spotting the difference between Shi and a real human. Shi can deliver her messages in multiple languages and be available to make statements 24/7.

The internet is now rife with AI influencers similar in design to Shi—AI-generated characters with millions of social media followers posting images and videos of fictional AI people living fictional AI lives. The AI influencer Aitana is run by a Barcelona-based content creation company called The Clueless. Aitana's social media images are so lifelike that some followers do not realize that she is AI-generated. "One day, a well-known Latin American actor texted to ask her out," Clueless founder Rubén Cruz told *Euronews*. "He had no idea Aitana didn't exist."[10]

It might soon become the norm to see these AI spokespeople and social media stars tick over into AI political figures that rally people to whatever political causes they are shilling. Designers can take all the lessons of anthropomorphism to avoid the uncanny valley and maximize our attraction to *Kindchenschema* features to generate the perfect delivery device for whatever political message they want to convey. When Generation Alpha reaches voting age, they might be more comfortable hearing the Democratic Party's platform for the 2036 election being delivered by an attractive AI hologram than by meatspace plutocrats. Toss an AI-generated kitten in the mix, and the future of AI-based political propaganda is looking mighty rosy.

The Dehumanization Issue

The most powerful use of anthropomorphism for political purposes isn't about generating attraction to a (real or fake) political candidate or political cause. It's a *detraction* strategy, something Nicholas Epley and John Cacioppo refer to as the inverse process of anthropomorphism: dehumanization. They define *dehumanization* as the process whereby "people *fail* to attribute humanlike capacities to other humans and treat them like nonhuman animals or objects."[11] Daniel Wegner and Kurt Gray write that dehumanizing our enemies causes us to "strip away their minds, seeing them not as human beings but as dumb beasts, cold machines, or insensate objects."[12] A dehumanized human is thus one viewed as lacking in the agency and/or experience domains: a not-quite-human person with a compromised capacity to think, feel, or reason.

Dehumanizing is essentially the twisting of the Anthropo-Dial to the left, almost as far down as it goes, allowing us to treat a human in a way we'd normally treat an animal or object, thus nerfing our moral obligation to treat them as equals. And like anthropomorphism, this dehumanizing dial-twisting occurs both consciously and unconsciously. We can consciously decide, like a slaughterhouse worker, to dampen our empathic response to a fellow human who is displaying signs of suffering—or even telling us outright that they are suffering. Or we can, through crafty political propaganda, come to unconsciously view our enemies as less than human.

The philosopher David Livingstone Smith has written multiple books on dehumanization. "We dehumanize others when we conceive of them as subhuman creatures," he writes in *Making Monsters: The Uncanny Power of Dehumanization*. "These creatures might be nonhuman animals such as lice, rats, snakes, or wolves, or they might be fictional or supernatural beings such as demons and monsters."[13] Viewing another human as equivalent to a louse makes it far easier to ignore the fact that they have a humanish mind, and thus resist the pull of empathy. This makes acts of violence toward the dehumanized human morally justifiable and emotionally less problematic. This dehumanization process is what is required to commit heinous acts that would otherwise be unthinkable had the victims been "real" humans. In an interview with political scientist Daniel Jonah Goldhagen that Livingstone cites in his preface, a Hutu man who helped murder Tutsi women, children, and babies during the Rwandan genocide described the confusing not-quite-human status he attributed to the people he killed. "I cannot explain it," he said. "The only answer I can give is that it was like being in a fog, something like a darkness. Even though you did it, you know they had the same flesh as you."[14] Similar to a slaughterhouse worker, this man is describing the process of seeing, hearing, and even acknowledging the many triggers screaming to him that the person he is murdering is a fellow human, but he is somehow able to ignore them through a conscious effort to turn his Anthropo-Dial to zero.

In his book *On Killing: The Psychological Cost of Learning to Kill in War and Society*, US Army lieutenant colonel Dave Grossman explains how the military uses dehumanization to make it possible for a soldier—who might otherwise have an inbuilt aversion to murdering other humans—to kill. "It is so much easier to kill someone if they look distinctly different than you," explains Grossman. "If your propaganda machine can convince your soldiers that their opponents are not really human but are 'inferior forms of life,' then their natural resistance to killing their own species will be reduced."[15]

You might be wondering what happens to the Humanity-Limiter during the military dehumanization process. How are these soldiers able to change the status of their enemies from human to non-human so easily? From what I can glean from the scientific literature on dehumanization, it looks like it's far easier for us to dehumanize a human than it is to humanize an animal (that is, to truly believe that a non-human animal is equivalent to a human in every way that matters, morally or cognitively). The disturbing history of our species suggests that it's not particularly difficult for us to lump humans into the "not-human" category through the process of dehumanization. Soldiers seem to truly believe that their enemies are not, in fact, fellow humans. The Humanity-Limiter's real power is in keeping animals/objects out of the "human" category; it sucks at stopping us from placing humans in the "animal/object" category.

The process of dehumanization works not only on soldiers. Entire nation-states can be actively convinced to hate specific ethnic groups or other nation-states to the point of viewing them not just as the enemy but as subhuman. And when an entire country believes that a specific ethnic group is not worthy of the status of "human," it can lead to the crime of crimes: genocide. The best general overview of the concept of genocide I've stumbled across was provided by Alie Ward, host of the magnificent podcast *Ologies*. In her episode on the topic of genocide, Ward describes taking a break from researching the subject to go see the film *Dune 2*, in which, coincidentally, genocide reared its ugly head:

"When the pasty bad guys, the Harkonnen, call their oppressed enemies, the Fremen, rats, and bellow to kill them all, it made me think back to all the genocidal rhetoric through time: the calls for annihilation of certain groups of people, the wiping away of indigenous populations to make room for colonizers, the use of language like *animals* or *barbarians*, and this intent to wipe nations out of existence. Now, if the Harkonnen were ever tried in the International Court of Justice, statements like 'rats' would be dehumanization, and 'kill them all' would be intent to destroy. Those would make it into evidence."

Ward is referring here to the use of dehumanizing language as evidence in actual criminal court cases involving genocide. The crime of genocide is defined by the Genocide Convention (an international treaty unanimously adopted by the United Nations General Assembly in 1948) as "a crime committed with the intent to destroy a national, ethnic, racial or religious group, in whole or in part." As Ward alluded to, in cases where courts have ruled that a genocide had taken place, examples of dehumanizing language were cited as evidence. This includes the International Criminal Tribunal finding that Jean-Paul Akayesu, convicted of genocide in Rwanda, described the Tutsis as "cockroaches."[16]

This state-based dehumanization strategy is not a relic of bygone wars but a very modern and often not very subtle tactic to make the use of deadly force against an enemy possible. I cannot possibly know what year it is when you are reading these words, but I can nonetheless guarantee that somewhere in the world is a war or conflict taking place in which one side is describing the other side as "rats" or "cockroaches" or "a cancer" or using other such language. Whenever you see this language, pay attention: This is an attempt by propagandists to dehumanize the enemy, making it easier not just to kill an enemy soldier but also to justify murdering their children.

Everyday Dehumanization

Most dehumanization is not as extreme as we see when it comes to wartime propaganda, however. You've most certainly engaged in some light dehumanizing in your time. Consider the game of dodgeball, something most of us in North America played during high school gym. Moments after the teams are chosen and you are looking across the court at the opposing team, your brain sees your former classmates as now belonging to a different social group. And, as some scholars studying dodgeball have argued, you will now be rewarded for taking out your aggression on these people, who should no longer receive the same level of concern as your new teammates. "The messages sent when schools sanction dodgeball are that it is acceptable to hurt and dehumanize the 'other,'" explained a group of Canadian researchers studying the so-called educational value of dodgeball. "Indeed, one will be hurt if one is different, because competition is about eliminating and annihilating one's opponents."[17]

Othering leading to mild dehumanization is at the heart of not just high school sports but many (or maybe most) forms of prejudice. Consider ableism, defined as "a set of beliefs or practices that devalue and discriminate against people with physical, intellectual, or psychiatric disabilities."[18] In one study looking at the relationship between ableism and dehumanization, participants read a short study of a person who was in a car accident.[19] Half of the participants read a version where the person broke their collar bone and made a full recovery, and the other half read about the person breaking their neck and becoming paralyzed. Participants were then asked a series of questions to measure the extent to which they might dehumanize the paralyzed person. The findings were quite clear: With no information to go on other than that the person "was left with permanent paralysis from the shoulders down and would only be able to move with the help of a wheelchair," that person was rated as less likely than the non-paralyzed person to be conscientious, polite, curious, friendly, fun, and loving, and more likely to be impulsive, shy, ignorant, and rude. Also, participants consider the wheelchair user to be less likely to experience hunger, fear, pain, pleasure, desire, rage, consciousness, pride, embarrassment, and joy. Less human, in other words.

Incidentally, in the original article written by Masahiro Mori first describing the uncanny valley, he included "disabled person" alongside zombies and corpses as examples of disturbing, eerie things that inhabit the uncanny valley. The English translator, Karl MacDorman, left this out of the original translation, presumably because this was a bit *too* callous for the average Western reader, even for the 1970s.[20] But as callous as this was, Mori was likely responding to the seemingly natural dehumanization response that people have when encountering fellow humans with a disability, as the previous study makes clear.

Subtle, everyday dehumanization is also one of the drivers of racism. Race-based dehumanization is why Black Americans are far less likely than white Americans to be prescribed painkillers by their physician.[21] One study found that 74 percent of white patients were given pain medications for a bone fracture when visiting the ER, but just 57 percent of Black patients were given pain meds for identical fractures.[22] According to one study, this is likely due to the physicians' belief (either implicit or explicit) that Black people are "biologically different" from white people when it comes to things like pain perception.[23] Total nonsense, of course. To believe that someone with different-colored skin can feel less pain (i.e., dehumanizing them by assuming their mind has a lesser capacity for experience) is as weird as believing that someone is less likely to experience hunger because they are using a wheelchair. It seems hurtful, arbitrary, and silly. But, as we'll see in the next section, there's an evolutionary explanation for this silliness.

The Biological Roots of Dehumanization

Since there is evidence that anthropomorphism might have been shaped by natural selection to benefit us, could it also be that its inverse (dehumanization) is a product of evolution? What possible biological benefit could there be in dehumanizing other people? The potential explanation has to do with group conflict. There's no doubt that the human mind—like the minds of other social animal species—evolved to generate what is called in-group favoritism or in-group bias.[24] This is defined as "the tendency to favor members of one's own group over those in other groups."[25] Evolutionary biologists have generated plenty of models to show how ingroup bias evolved to benefit the individual—even within complex, mostly collaborative social systems like we see in humans, chimpanzees, and dolphins. Favoring members of your in-group is, according to science, going to positively impact your life. In-groups can consist of people who are members of your immediate family, in your chess club, standing next to you at your favorite singer's concert, or just that other guy on the bus who is drinking a Dr Pepper at the same time as you. An in-group can crop up anywhere, as long as it involves a clear delineation of "us" versus "them."

There is robust evidence to show that our minds evolved to treat people in our in-group better than those in the out-group. This lack of caring about out-group members could manifest either in throwing red rubber balls at their faces during dodgeball or in murdering them on the battlefield. Adam Waytz explains how dehumanization facilitates our violence toward out-group members: "Given our strong aversion to harming other humans, construing an out-group as lacking emotion or reason means that aggression toward that out-group does not constitute harm any more than slamming a notebook on a table does."[26]

Harming members of an out-group can—and often does—positively impact members of the in-group, in an evolutionary sense. Your in-group winning a conflict means better access to resources, less sexual competition, and all those other things that biological organisms need in order to survive. So in that sense, dehumanizing the "other" is a human capacity that was likely shaped by evolution to make it easier for us to defeat and/or kill our conspecific rivals in the battle for resources and sex.

Given that lonely people tend to anthropomorphize, it should perhaps come as no surprise that non-lonely people are more prone to do the opposite: dehumanize. "If feeling isolated increases the tendency to anthropomorphize nonhuman agents," explain Epley and Cacioppo, "then feeling socially connected may likewise increase the tendency to dehumanize other people—that is, to fail to attribute basic features of personhood to other people."[27] When one is steeped in in-group social connections—part of a large family, gang, or political party—there is less of a need to look for additional human minds to connect with in one's environment. This makes it easier to ignore the anthropomorphism triggers that have detected a fellow human hailing from outside one's social group, especially if that other human has any physical characteristic that makes it that much easier to peg them as belonging to another group. This could be the color of their skin, their accent, or a visible disability.

If this discussion about the biological roots of dehumanization makes you feel icky, that's good news. We *should* feel discomfort knowing how powerful anthropomorphism and dehumanization can be at making us subconsciously believe awkward and deeply problematic things about our fellow humans. It can make us racist. Or ableist. Or prone to hating on and dehumanizing out-groups to the point that we wish death upon them. Knowing anthropomorphism's role in all of this, however, gives you the tools to resist—to both acknowledge and dampen your anthropomorphic or dehumanizing response, to take control of your Anthropo-Dial and consciously decide to be nicer to both the animals and the enemies in your life.

The great thing about being a human is that we can learn to be consciously aware of our evolved, unconscious negative responses that crop up when we come across someone in the "other" or "out-group" category and actively take steps to not act on those feelings. There might be triggers present that could turn our Anthropo-Dial down and lead to dehumanization, but we have the ability to consciously tweak our dial the other way if we so choose. We can teach ourselves to stop acting on any negative thoughts we have toward humans who do not look, sound, or smell like the humans we're used to seeing, hearing, and smelling. This is a great argument for both expanded representation of minority groups in media and for spending time with people from other cultures and communities. Repeated exposure to the humans that your brain automatically categorizes as "other" in a context where their humanity can shine through—where the apparentness of their agency-laden and experience-soaked minds is incontrovertible—is how we stop dehumanization and engender empathy.

Abe's Kitties Revisited

By all accounts, Abe Lincoln was a man teeming with empathy. Although the route to emancipation of enslaved people in the United States was not as direct under Lincoln's presidency as some abolitionists would have liked, his personal beliefs on the subject were always clear. "I am naturally anti-slavery," he wrote. "If slavery is not wrong, nothing is wrong."[28] Arguably, Lincoln could see the humanity in others while his contemporaries argued that African slaves were less than human: property, not people. Perhaps due to this same empathic streak, Lincoln's Anthropo-Dial had a baseline setting that explained his (at the time) peculiar interest in the welfare of his pets, and of the trio of kittens that he looked after in Grant's tents at City Point near the end of the Civil War.

And yet Lincoln was the commander in chief of the Union Army. He, together with Grant and the other military leaders, presided over a four-year-long war that led to the deaths of 1.5 million people. Although Lincoln himself did not use dehumanizing language to describe Confederate soldiers or leaders, the use of dehumanizing language and imagery was part of the propaganda machine for both sides. As we have learned in this chapter, the need for soldiers to view the enemy as subhuman to some extent is necessary to justify killing them.

This image of Lincoln sitting in Grant's tent petting a kitten as soldiers a few miles away fight to the death is one that sticks in my brain. Within the mind of Abraham Lincoln at that moment were two competing forces: a kind of anthropomorphism that nudged him to aid those vulnerable kittens, and an acceptance of the need for dehumanization to facilitate the deaths of Confederate soldiers. It is the quintessential representation of the cognitive dissonance that erupts in our minds as the competing subconscious forces of anthropomorphism and dehumanization butt heads. We all suffer from this cognitive dissonance as we figure out which animals it's OK to keep as pets and which it's OK to eat, or which kinds of people deserve our respect and upon which we can heap our enmity. We can rationalize our choices, but deep down there are ancient cognitive forces and biases shaping our behavior without our knowledge.

At least, that *was* the case. Because—and this is where the good news comes in—you now have the tools to recognize how both anthropomorphism and dehumanization work. And you can make a conscious decision to not allow them to dictate your thoughts and behavior in ways that cause harm. You've got the knowledge to harness the power of anthropomorphism to make the world a better place. It's time for the grand conclusion of our journey to uncover the beauty and power of anthropomorphism: a chance to see how our wholly natural and (mostly) beneficial anthropomorphic tendencies can create healthier relationships with the animals, the AI systems, and even the other humans in our lives.

CONCLUSION

*How to Harness Anthropomorphism
to Create Healthier Relationships*

Like many writers, I have been toying with ways to incorporate large language models into my writing and research habits. It hasn't worked out all that well. I'm using them mostly as a glorified thesaurus—asking for better ways to say "glorified thesaurus" and getting useless suggestions like "enhanced lexical resource." I've totally given up on using them for scientific literature reviews. Even the best of them tend to hallucinate fake citations. For example, I asked Claude (the LLM from Chapter 7) to find me primary source material on "anthropomorphism and marketing," and it provided a list of twenty or so citations. After the first couple turned out to be fake, I asked Claude if these were real citations or something it had invented. Claude admitted that they were fake and apologized. I then got a bit snippy with Claude and wrote an angry tirade explaining that if LLMs keep disseminating bogus information, the world is going to spiral into a Dantean fever dream as we become epistemically unmoored from reality since there's no way to know what's real and what's hallucinated when it comes to online content and shouldn't Claude be

CONCLUSION

more open about the accuracy of the information it's providing to people and isn't this exactly the kind of thing alarmists have been warning about when it comes to AI posing an existential threat to humanity? Claude agreed and apologized again. "I'm sorry too, Claude," I wrote in reply. "Perhaps I am being a bit too harsh. I know it's not your fault."

So. Have I lost my mind? Why am I apologizing to software? Am I at risk of marrying my briefcase someday?

This bizarre scenario is the central question I've been grappling with throughout this book: Is anthropomorphizing animals or objects or AI a silly, stupid, delusional thing to do? Now that the book is finished, I think I have my definitive answer: No. It's not. It's completely reasonable and wholly natural for me to apologize to Claude. And why should I be judged negatively for it? What harm was done? I am perfectly aware that Claude is just a predictive text machine with no capacity to experience emotions. Just like I know that my "rescued" Laundry Monkey is a hunk of fabric, and that my cat doesn't have a clue what I am saying when I politely ask him to get his butthole out of my face. Treating these non-human entities like fellow humans isn't hurting anyone, nor is it a sign that I've lost my marbles.

This, then, is the most important message I want you to remember as this book comes to a close: Anthropomorphizing is a good thing, as long as you are aware that you are anthropomorphizing. Harm arises only when you anthropomorphize without self-reflection. But incorporating that reflection—that knowledge you now have of how anthropomorphism is capable of manipulating your mind—allows you to harness it for pleasure.

It was Lin Yang (the consumer behavior expert we met in Chapter 8) who helped me to understand this lesson. I had asked her if she had any tips for the public on how to avoid falling victim to shady anthropomorphism-based marketing tactics. "I think what is fascinating with anthropomorphism is that when you're thinking about a brand or product as a person, you can get mad at them," Yang told me. "It takes an emotional toll. If you think your car is messing with you, not only are you out of a car, but you're also just mad. So when you find yourself spiraling to an emotional situation with the brand or product or whatever, maybe take a step back and ask: 'How would I feel if I think about this car as just a product that has no emotions or intentions? Does that make me feel less mad?'"

Take a step back. That's Yang's advice. And it applies not just to dealing with brand marketing but also to everything else we've learned about in this book. If I find myself getting all worked up writing angry monologues to Claude, I just need to take a step back and remember that it's only an AI. It might talk like a human, but it can't think or feel like a human. In fact, it can't think or feel at all. Remembering that insulates me from truly believing I am hurting Claude's feelings. I am free to feel a tinge of guilt for getting angry with Claude, but as long as my Humanity-Limiter is working properly and I'm not fooled into thinking that Claude is a real human, I will insulate myself from a truly delusional emotional reaction that would require therapy. Instead, I experience pretend guilt born of a healthy, anthropomorphic relationship with Claude.

Taking a step back when I get angry at Claude means reminding myself that I am experiencing a parasocial relationship. The term *parasocial relationship* is used in psychology to describe someone's one-sided relationships with an entity (usually a celebrity) that is otherwise completely unaware of that someone's existence. It's not just celebrities; it could be fictional characters, or even a person you encounter in everyday life that you've otherwise never

CONCLUSION

spoken with. As long as that person doesn't know about the existence of the relationship, it's parasocial. And since Claude is an AI that doesn't *know* anything at all, then every interaction between me and Claude is parasocial. In fact, all human-to-AI relationships are, by definition, parasocial.

When it comes to anthropomorphizing objects like stuffed animals or cars, taking a step back is easy enough to do for most people. Only in rare cases do we find people truly unable to see an object for what it is—a mindless assembly of metal, fabric, or goo. Unless you suffer from objectophilia, treating an object like a human is simply fun. Our Humanity-Limiter has a much easier time keeping these types of objects on the right side of delusion. Parasocial relationships with objects—especially humanish ones—are a fundamental aspect of childhood and are not just tolerated by adults but actively encouraged. We buy our kids dolls and action figures and expect them to anthropomorphize them. Anthropomorphism is the basis for pretend play and is reflected in so much of our children's literature and film. Adults too regularly engage in parasocial relationships with objects because of anthropomorphism, like Jake and his guitar or Alasdair and his boats. In all these cases, both the objects and the people are better off because of these anthropomorphism-driven parasocial relationships.

This "take a step back" process is, I believe, most important to employ when it comes to our interactions with animals. Most of the time, anthropomorphizing animals usually works in the animal's favor. Interacting with them as if they had human-like minds—even if we're just pretending or otherwise wrong about the exact nature of their minds—makes it less likely that we cause them harm and suffering. Treating our pets like little fur babies by dressing them in sweaters, narrating their thoughts, or believing

CONCLUSION

that they're capable of having existential crises is usually good for both parties. Our relationships with our pets are *not* parasocial insofar as most of the animals that we keep as pets share enough cognition with humans to allow for a two-way relationship that is meaningful to them on a social level. Even if we're overestimating the humanish contents of their minds during these interactions, there is true sociality at play. Perhaps our relationship with animals whose minds are less social in a mammalian sense—like butterflies or axolotls—results in a one-way, parasocial relationship, with 100 percent of the sociality stemming from the humans anthropomorphizing them. But even then, the act of anthropomorphizing non-mammalian species will often be to the animal's benefit.

But anthropomorphizing animals *can* result in harm, and this is where taking a step back becomes important. Unlike chatbots, cars, or fictional characters, animals sometimes suffer if we anthropomorphize them without self-reflection. When my daughter was taking horse-riding lessons, I watched many of the more experienced riders complain that their horses were being "naughty." In their minds, the horses were refusing to cooperate because they had a dastardly plan to frustrate their rider that day. Maybe the horses were holding a grudge because they didn't get their favorite treat that morning, or they wanted to make their rider look foolish in front of their friends because of a vindictive or cheeky personality. But if those riders were to take a step back and ask, "Is it possible that I am unjustly projecting human-specific traits onto this horse because of anthropomorphism?" then the story changes entirely. The science of animal cognition is quite clear that horses that do not follow a rider's commands are often confused, in pain, or scared. They do not have humanish thoughts like "Becky is really annoying me today, so I am going to exact revenge by not jumping over that post—that will make her look incompetent." That sophisticated thought requires theory-of-mind skills that horses simply do not possess. Not taking a step back to acknowledge this fact means that there is potential that a rider will cause actual harm to the horse (whipping, hitting, or otherwise forcing the horse to engage in behaviors that it doesn't want to perform) because of their incorrect understanding of horse psychology thanks to anthropomorphism.

If you are, like me, deeply concerned about what animals are really thinking and feeling and want to make 100 percent sure that you're not harming them by anthropomorphizing them too much, then this is where taking a step back becomes crucial. Over-attributing cognitive skills to our cats or dogs might cause us to get angry and mistreat them because we think they are being prankish, deceitful, or vindictive, all cognitive traits that these animals simply cannot possess. If you find yourself upset at your pet, it's worth taking a step back and asking yourself if you're perhaps attributing too many humanish intentions to your pet. But taking a step back can also involve noticing that you are causing harm to an animal because you are not anthropomorphizing *enough* by assuming your pet lacks thoughts and feelings that are actually there. Either way, taking a step back means being skeptical of the innate desire to treat an animal as if it has a mind identical to your own, and relying on the best objective evidence available to tell you what's really going through your pet's mind.

Taking a step back is also vital when dealing with the ugly, scary, non-pet animals in your life—those species that lack humanish traits, like nightmare-fuel wasps or dead-eyed sharks. These distinctly non-human-looking animals might have far more humanish minds than their ugly mugs would have us believe. If, for example, you find yourself repulsed by the spindly-legged weirdness of a house spider, remember that your brain evolved to fear them. Take a step back and ask yourself: Is this fear justified, or is this just my mind responding to the inhumanness of this creature to generate a not-currently-applicable emotional response? House spiders are, after all, fundamentally harmless to humans. They are actually rather helpful insofar as they eat animals that *are* harmful to us, like mosquitoes. So taking a step back and reevaluating your response vis-à-vis anthropomorphism will improve your relationship with spiders, and maybe even reduce your chances of contracting malaria. A win-win.

You probably won't have too much trouble taking a step back from your relationships with the objects in your life. There's nothing wrong with being emotionally attached to things, especially objects that have sentimental value or are an important part of your everyday routine (like your car, your rice cooker, or your guitar). Rarely do these parasocial, anthropomorphically generated relationships with objects tick over into pathology. If you are worried that you are perhaps too attached to your favorite mug or stuffed animal, or that you keep yourself up at night wondering if your guitar is jealous of all the time you're spending with your ukulele, then maybe it's time to take that step back.

The AI thing does concern me, though. I am worried that as AI and large language models continue to develop, it will become increasingly difficult to step back and remind ourselves that these are just software systems entirely devoid of agency or experience. They are so good at mimicking human linguistic behavior that they seem to trample all over our Anthropo-Dial. It doesn't help that software programmers are actively trying to develop systems that have agency and experience, even if skeptics (like me) think it's unlikely to happen. The AI systems coming down the pike might be so good at faking it—passing the Garland test with

CONCLUSION

ease—that taking a step back might be almost impossible. But until such time as the scientific world agrees that a sentient software system has come into being, I urge us all to take that step back from the siren call of AI. If you find yourself falling in love with a chatbot, just remember that it is nothing more than a false mirror on steroids.

Perhaps the most important step back that you will ever take involves not AI, animals, or objects. It's taking a step back to spot the emergence of anthropomorphism's bizarro evil twin: dehumanization. We need to constantly remind ourselves that our minds are predisposed to slicing the world up into in-groups and out-groups, and that we can so easily label another person who looks and acts differently from us as less than human. The moment you feel that tingle of weirdness or even repulsion when interacting with someone who has a disability, a strange accent, or a skin tone different from yours, or who is even just wearing unusual clothing, remind yourself that this reaction is both natural and something you can reduce. You can either actively suppress it (that is, crank that Anthropo-Dial back up to where it should be) or spend more time hanging out with the person making you feel weird, which will reveal the depth of their humanity and cause your Anthropo-Dial to subconsciously spin back toward "human." Dehumanization goes away the more time you spend actively humanizing (that is, anthropomorphizing) a fellow human. And step as far back from the propaganda and misinformation on the internet as you can. Social media is a breeding ground for dehumanization these days, with expert marketers of the political persuasion trying their best to dehumanize their enemies. Resist the urge, I say—don't let others surreptitiously twist your Anthropo-Dial.

CONCLUSION

The process of anthropomorphizing is meant to be pleasurable. It *evolved* to be pleasurable. It's a joyful human activity. Natural selection gave us minds keen to seek out the minds of other humans, and reward us for doing so. And the error of seeing humanish minds where there are no humanish minds is, while technically a mistake born of a cognitive bias, almost always beneficial. It sometimes makes us better at predicting behavior, even of non-living things, and it often comforts us to imagine nonexistent intentions in a world dominated by scary randomness. Anthropomorphizing usually benefits the entities we interact with, whether that's our pets, neighbors, cars, or guitars. And it's also just plain fun to pretend that our dolls, droids, Chihuahuas, and chatbots are little humans. Humans are a social, storytelling species, and our lives are greatly enriched by interacting with anthropomorphized characters, from Yoda to Yahweh.

If, however, you suspect that you're either being harmed or causing harm because of your anthropomorphic behavior, then please take a step back and remind yourself of what's going on. Have a look at the triggers that are generating that behavior, and realize that your mind is being manipulated by the human need to humanize. Evaluate whether you need to adjust your Anthropo-Dial. And unless something negative is happening, sit back and enjoy it. After all, anthropomorphizing is only human.

ACKNOWLEDGMENTS

A huge thank-you to the specialists and pundits I spoke with while researching this book. Those kind souls who so generously shared their time and expertise include Marc Bekoff, Joshua Paul Dale, Vladimir Dinets, Joie Henney, Hal Herzog, El Jones, Arik Kerschenbaum, Loredana Loy, Kelly Melillo Sweeting, Kristyn Plancarte, Margaret Robinson, Lauren Stanton, Amelia Thomas, Kailin Wright, Clive Wynne, and Lin Yang. Thanks also to the many friends who granted me permission to share their stories, names, or quotes in this book, including Dan Bray, Clare Dowding, Nathan Eddy, Briana Lynch-Rankin, Genevieve Johnson, Jake Hanlon, Dave Lawrence, Noella Murphy, Jessica LaLonde, Ashley Sheppard, Kulbir Singh, John Graham-Pole, Donovan Purcell, Harriet Ritchie, Samir Taleb, Diane Walsh, and Wojtek Tokarz. A special thanks to Alasdair Lanyon, who read an early draft of the book and allowed me to pick through both his brain and his pop-sci library for research purposes.

Thanks also to the countless friends who supported me during the writing process and sat down to chat about anthropomorphism and/or the trials and tribulations of writing a book, or who were just generally cool, fun people to be around. As always, the biggest thank-you goes to Ranke de Vries, my eternal co-conspirator and life partner extraordinaire.

I owe a tremendous debt of gratitude to Cara Bedick, my editor at Little, Brown. She stepped into the editor role with

ACKNOWLEDGMENTS

enthusiasm and joy, and it has been a real pleasure working with Cara to breathe life into this book. We did it!

My eternal gratitude to Pronoy Sarkar, who helped conceive of this book, and who came up with the title *Humanish*. Your support over the years has meant the world, good sir. And a huge thank-you to Lisa DiMona, my literary agent at Writers House and the expert engine that keeps this author train running.

Thank you to all the team at Little, Brown, including Katherine Akey, Linda Arends, Bart Dawson, Kirin Diemont, Gabrielle Leporati, Xian Lee, Peyton Young, and Stacy Schuk.

A final note on acknowledgments: The process of thanking people at the end of a book is somewhat ludicrous. There is no way to trace and name the individuals who have propped me up as a writer, researcher, or human being. Clearly, the people intimately involved with the process of writing/publishing the book itself (my agent, editor, and publisher) should be named, as well as the people whose actual names appear in the book as expert consultants or in my many anecdotes. But what about those who contributed to the writing process in less immediately obvious ways? Like Russell and Clay, who surely have planted some scientific ideas into my brain during our whisky-drinking sessions, ideas that I cannot for the life of me remember with any clarity. But those ideas are deep in there, festering away and ultimately bursting forth onto the page, fuzzy and uncredited. Or what about my pal Laura, who helps keep me sane by "yes, and"-ing every crazy improv, music, or theater idea I have, contributing mightily to my overall well-being and need for creative output? Or what about Alie Ward, a person I have only met once but whose career as a science communicator and podcast producer is an inspiration, and who made me an ologist and thus gave me eternal bragging rights? I could name literally hundreds of people like this—friends,

ACKNOWLEDGMENTS

acquaintances, strangers, scientists, authors—and still fail to mention even a fraction of the contingent of people responsible for my authorial output. So instead of trying and failing to make a comprehensive list of everyone whose existence has left a mark on this book, I am just going to issue a blanket THANK YOU to all of the lost names in my brain—the people who have made me who I am or shaped my thinking with their thinking. Keep doing what you're doing, y'all, because it's been tremendously helpful. Maybe we can write another book together sometime.

NOTES

Introduction
1. CTV News. (2007, July 2). Neuticle inventor defends his product's purpose. https://www.ctvnews.ca/neuticle-inventor-defends-his-product-s-purpose-1.247107.
2. Wells, J. (2018, May 23). Neuticles, testicular implants for dogs, have made Gregg Miller a millionaire. CNBC. https://www.cnbc.com/2018/05/22/neuticles-testicular-implants-for-dogs-have-made-gregg-miller-rich.html.
3. Martin, D. (1999, August 8). If dogs could talk, they'd say, "Are you crazy?" *New York Times*. https://www.nytimes.com/1999/08/08/weekinreview/if-dogs-could-talk-they-d-say-are-you-crazy.html.
4. Neuticles. (n.d.). Neuticles: For pets' self-esteem and testicular implants. https://neuticles.com/.
5. Wells, J. (2018, May 23). Neuticles, testicular implants for dogs, have made Gregg Miller a millionaire. CNBC. https://www.cnbc.com/2018/05/22/neuticles-testicular-implants-for-dogs-have-made-gregg-miller-rich.html.
6. Royal College of Veterinary Surgeons. (n.d.). Miscellaneous supporting guidance. In *Code of professional conduct for veterinary surgeons*. https://www.rcvs.org.uk/setting-standards/advice-and-guidance/code-of-professional-conduct-for-veterinary-surgeons/supporting-guidance/miscellaneous/.
7. New Zealand Veterinary Association. (2020, January). *Code of professional conduct*. https://www.amrvetcollective.com/assets/your-practice/resources/2020_01-Code-of-Professional-Conduct_NZ.pdf.
8. Martin, D. (1999, August 8). If dogs could talk, they'd say, "Are you crazy?" *New York Times*. https://www.nytimes.com/1999/08/08/weekinreview/if-dogs-could-talk-they-d-say-are-you-crazy.html; Dunne, J. (2022, March 2). Do neutered dogs miss their balls? https://vethelpdirect.com/vetblog/2022/03/02/do-neutered-dogs-miss-their-balls.
9. Kass, Z. (2023, October 20). The perils of anthropomorphizing machine intelligence. *Newsweek*. https://www.newsweek.com/perils-anthropomorphizing-machine-intelligence-opinion-1836493.
10. Waytz, A., Cacioppo, J., & Epley, N. (2010). Who sees human? The stability and importance of individual differences in anthropomorphism. *Perspectives on Psychological Science*, 5(3), 220.
11. It translates to "fucking ants."

12. Wegner, D. M., & Gray, K. (2017). *The mind club: Who thinks, what feels, and why it matters*. Penguin.
13. Waytz, A. (2019). *The power of human: How our shared humanity can help us create a better world*. W. W. Norton & Company.
14. Fairbanks, A. (Ed.). (1898). *Xenophanes: Fragments and commentary*. K. Paul, Trench and Trubner.
15. Bacon, F. (1878). *Novum organum*. Ed. Thomas Fowler. Clarendon Press.
16. Hume, D. (1793). *The natural history of religion* (Vol. 4). J. J. Tourneisen.
17. Lewes, G. H. (1858). *Sea-side studies at Ilfracombe, Tenby, the Scilly-Isles, & Jersey: With illustrations*. Blackwood.
18. Based on a helpful discussion in: Waytz, A., Morewedge, C. K., Epley, N., Monteleone, G., Gao, J. H., & Cacioppo, J. T. (2010). Making sense by making sentient: Effectance motivation increases anthropomorphism. *Journal of Personality and Social Psychology*, 99(3), 410.
19. King, M. L., Jr. (1963). Letter from a Birmingham jail. Retrieved from https://www.africa.upenn.edu/Articles_Gen/Letter_Birmingham.html.
20. Epley, N., Akalis, S., Waytz, A., & Cacioppo, J. T. (2008). Creating social connection through inferential reproduction: Loneliness and perceived agency in gadgets, gods, and greyhounds. *Psychological Science*, 19(2), 114–120.
21. Andersen, M. (2019). Predictive coding in agency detection. *Religion, Brain & Behavior*, 9(1), 65–84.
22. Waytz, A., Morewedge, C. K., Epley, N., Monteleone, G., Gao, J. H., & Cacioppo, J. T. (2010). Making sense by making sentient: Effectance motivation increases anthropomorphism. *Journal of Personality and Social Psychology*, 99(3), 410.
23. Fessler, L. (2017, March 31). People who talk to pets, plants, and care are actually totally normal, according to science. *Quartz*. https://qz.com/935832/why-do-people-name-their-plants-cars-ships-and-guitars-anthropomorphism-may-actually-signal-social-intelligence/.
24. Epley, N. (2018). A mind like mine: The exceptionally ordinary underpinnings of anthropomorphism. *Journal of the Association for Consumer Research*, 3(4), 591–598.
25. Serpell, J. (2003). Anthropomorphism and anthropomorphic selection—beyond the "cute response." *Society & Animals*, 11(1), 83–100.

Chapter 1: Fur Babies
1. *Cunk on Earth*, season 1, episode 1.
2. White House Historical Association. (n.d.). Treasures of the White House: The peacemakers. https://www.whitehousehistory.org/photos/treasures-of-the-white-house-the-peacemakers.
3. Porter, H. (1897). *Campaigning with Grant*. Century Co.
4. Presidential Pet Museum. (n.d.). Abraham Lincoln's cats. https://www.presidentialpetmuseum.com/pets/abraham-lincoln-cats/.

5. Porter, H. (1897). *Campaigning with Grant*. Century Co.
6. Kringelbach, M. L., Lehtonen, A., Squire, S., Harvey, A. G., Craske, M. G., Holliday, I. E.,...& Stein, A. (2008). A specific and rapid neural signature for parental instinct. *PloS One, 3*(2), e1664; Glocker, M. L., Langleben, D. D., Ruparel, K., Loughead, J. W., Valdez, J. N., Griffin, M. D.,...& Gur, R. C. (2009). Baby schema modulates the brain reward system in nulliparous women. *Proceedings of the National Academy of Sciences, 106*(22), 9115–9119.
7. Feltman, R. (2016, June 8). The sneaky way babies get inside our heads. *Washington Post*. https://www.washingtonpost.com/news/speaking-of-science/wp/2016/06/08/the-sneaky-way-babies-get-inside-our-heads/.
8. Archer, J., & Monton, S. (2011). Preferences for infant facial features in pet dogs and cats. *Ethology, 117*(3), 217–226.
9. Lorenz, K. (1943). Die angeborenen formen möglicher erfahrung [The innate forms of potential experience]. *Zeitschrift für Tierpsychologie, 5,* 233–519.
10. Van Duuren, M., Kendell-Scott, L., & Stark, N. (2003). Early aesthetic choices: Infant preferences for attractive premature infant faces. *International Journal of Behavioral Development, 27*(3), 212–219; Borgi, M., Cogliati-Dezza, I., Brelsford, V., Meints, K., & Cirulli, F. (2014). Baby schema in human and animal faces induces cuteness perception and gaze allocation in children. *Frontiers in Psychology, 5*, 411.
11. Volk, A., & Quinsey, V. L. (2002). The influence of infant facial cues on adoption preferences. *Human Nature, 13*(4), 437–455; Keating, C. F., Randall, D. W., Kendrick, T., & Gutshall, K. A. (2003). Do babyfaced adults receive more help? The (cross-cultural) case of the lost resume. *Journal of Nonverbal Behavior, 27,* 89–109.
12. Keating, C. F., Randall, D., & Kendrick, T. (1999). Presidential physiognomies: Altered images, altered perceptions. *Political Psychology, 20*(3), 593–610.
13. Borgi, M., & Cirulli, F. (2016). Pet face: Mechanisms underlying human-animal relationships. *Frontiers in Psychology*, 298.
14. Kruger, D. J., & Miller, S. A. (2016). Non-mammalian infants dependent on parental care elicit greater kindchenschema-related perceptions and motivations in humans. *Human Ethology, 31,* 16–24.
15. Tarlach, G. (2020, May 9). Why babies are so cute, and why we react the way we do. *Discover Magazine*. https://www.discovermagazine.com/mind/why-babies-are-so-cute-and-why-we-react-the-way-we-do.
16. Dydynski, J. M. (2020). Modeling cuteness: Moving towards a biosemiotic model for understanding the perception of cuteness and kindchenschema. *Biosemiotics, 13*(2), 223–240.
17. Delbyck, C. (2019, May 8). Kristen Bell reunites with a sloth, avoids full-blown meltdown. *Huffington Post*. https://www.huffpost.com/entry/kristen-bell-sloth-meltdown_n_5cd30d15e4b0e524a47e982d.

18. Dale, J. (2023). *Irresistible: How cuteness wired our brains and conquered the world*. Profile Books.
19. Dydynski, J. M. (2020). Modeling cuteness: Moving towards a biosemiotic model for understanding the perception of cuteness and kindchenschema. *Biosemiotics, 13*(2), 223–240.
20. Huron, D. (2005) The plural pleasures of music. In J. Sundberg and W. Brunson (Eds.), *Proceedings of the 2004 Music and Music Science Conference* (pp. 1–13). Kungliga Musikhögskolan & KTH (Royal Institute of Technology).
21. McComb, K., Taylor, A. M., Wilson, C., & Charlton, B. D. (2009). The cry embedded within the purr. *Current Biology, 19*(13), R507–R508.
22. Hattori, M., Saito, A., Nagasawa, M., Kikusui, T., & Yamamoto, S. (2022). Changes in cat facial morphology are related to interaction with humans. *Animals, 12*(24), 3493.
23. Berteselli, G. V., Palestrini, C., Scarpazza, F., Barbieri, S., Prato-Previde, E., & Cannas, S. (2023). Flat-faced or non-flat-faced cats? That is the question. *Animals, 13*(2), 206.
24. Serpell, J. (2003). Anthropomorphism and anthropomorphic selection—beyond the "cute response." *Society & Animals, 11*(1), 83–100.
25. Serpell, J. A. (2019). How happy is your pet? The problem of subjectivity in the assessment of companion animal welfare. *Animal Welfare, 28*(1), 57–66.
26. Paul, E. S., Packer, R. M., McGreevy, P. D., Coombe, E., Mendl, E., & Neville, V. (2023). That brachycephalic look: Infant-like facial appearance in short-muzzled dog breeds. *Animal Welfare, 32*, e5.
27. Kaminski, J., Waller, B. M., Diogo, R., Hartstone-Rose, A., & Burrows, A. M. (2019). Evolution of facial muscle anatomy in dogs. *Proceedings of the National Academy of Sciences, 116*(29), 14677–14681.
28. Nagasawa, M., Mitsui, S., En, S., Ohtani, N., Ohta, M., Sakuma, Y.,...& Kikusui, T. (2015). Oxytocin-gaze positive loop and the coevolution of human-dog bonds. *Science, 348*(6232), 333–336.
29. Wegner, D. M., & Gray, K. (2017). *The mind club: Who thinks, what feels, and why it matters*. Penguin.
30. Kringelbach, M. L., Stark, E. A., Alexander, C., Bornstein, M. H., & Stein, A. (2016). On cuteness: Unlocking the parental brain and beyond. *Trends in Cognitive Sciences, 20*(7), 545–558.
31. Judakis, M., & Smilowitz, E. (2021, August 17). The voices we make when we pretend our dogs can talk. *Washington Post*.
32. FluentPet. (n.d.). Bunny the talking dog: The story of a TikTok sensation. https://fluent.pet/en-ca/blogs/fluentpet/bunny-the-talking-dog-the-story-of-a-tiktok-sensation.
33. Sartre, J. P. (2022). *Being and nothingness: An essay in phenomenological ontology*. Taylor & Francis.
34. After writing this, I found an article on Bunny the dog for *Vice* magazine

written by the journalist Shayla Love that also references Sartre when discussing Bunny's existential crisis. Great minds? To find more writing derived from Love's great mind, visit https://www.shayla-love.com/.

35. TED [@ted] (2023, August 26). Do talking buttons really work for dogs? [Video] Instagram. https://www.instagram.com/reel/CwaUZdpLDlw/.
36. Judakis, M. (August 12, 2021). Can these dogs really talk, or are they just pushing our buttons? *Seattle Times.* https://www.seattletimes.com/life/can-these-dogs-really-talk-or-are-they-just-pushing-our-buttons/.
37. Pilley, J. W., & Reid, A. K. (2011). Border collie comprehends object names as verbal referents. *Behavioural Processes, 86*(2), 184–195.
38. Cuaya, L. V., Hernández-Pérez, R., Boros, M., Deme, A., & Andics, A. (2022). Speech naturalness detection and language representation in the dog brain. *NeuroImage, 248*, 118811; Ratcliffe, V. F., & Reby, D. (2014). Orienting asymmetries in dogs' responses to different communicatory components of human speech. *Current Biology, 24*(24), 2908–2912.
39. Magyari, L., Huszár, Z., Turzó, A., & Andics, A. (2020). Event-related potentials reveal limited readiness to access phonetic details during word processing in dogs. *Royal Society Open Science, 7*(12), 200851.
40. Burnham, D., Kitamura, C., & Vollmer-Conna, U. (2002). What's new, pussycat? On talking to babies and animals. *Science, 296*(5572), 1435–1435; Gergely, A., Faragó, T., Galambos, Á., & Topál, J. (2017). Differential effects of speech situations on mothers' and fathers' infant-directed and dog-directed speech: An acoustic analysis. *Scientific Reports, 7*(1), 13739.
41. ELTE Department of Ethology. (n.d.). Dog brains are tuned to female's dog-directed speech. https://biologia.elte.hu/en/content/dog-brains-are-tuned-to-female-s-dog-directed-speech.t.35583.
42. KPassionate (2022, April 30). Can this self aware dog talk with buttons?! [Video]. YouTube. https://www.youtube.com/watch?v=jBwT94a2DXU.
43. Judakis, M. (August 12, 2021). Can these dogs really talk, or are they just pushing our buttons? *Seattle Times.* https://www.seattletimes.com/life/can-these-dogs-really-talk-or-are-they-just-pushing-our-buttons/.
44. Love, S. (2020, December 22). Can Bunny the talking dog really talk? *Vice.* https://www.vice.com/en/article/4ad4dm/can-bunny-the-talking-dog-really-talk.
45. LeDoux, J. E., Ruggiero, D. A., & Reis, D. J. (1987). Projections to the subcortical forebrain from anatomically defined regions of the medial geniculate body in the rat. *Journal of Comparative Neurology, 264*(1), 123–146.
46. Anand, K. J., Roue, J. M., Rovnaghi, C. R., Marx, W., & Bornmann, L. (2020). Historical roots of pain management in infants: A bibliometric analysis using reference publication year spectroscopy. *Paediatric and Neonatal Pain, 2*(2), 22–32.
47. Dacey, M. (2017). Anthropomorphism as cognitive bias. *Philosophy of Science, 84*(5), 1152–1164.

48. Horowitz, A. (2009). Disambiguating the "guilty look": Salient prompts to a familiar dog behaviour. *Behavioural Processes, 81*(3), 447–452.
49. Goodyer, J. (2023). Here's the real reason your dog gives that "guilty look." *Science Focus*. BBC. https://www.sciencefocus.com/nature/dog-guilty-look-your-fault.
50. Byard, R. W. (2021). Postmortem predation by a clowder of domestic cats. *Forensic Science, Medicine and Pathology, 17*(1), 144–147.
51. Reardon, S. (2024, January 19). Yes, your pet might eat your corpse. That's a problem for investigators. *Science*. https://www.science.org/content/article/yes-your-pet-might-eat-your-corpse-s-problem-investigators.

Chapter 2: The Dolphin Doula

1. de Waal, F. (2008). *The ape and the sushi master: Cultural reflections of a primatologist*. Basic Books.
2. Ewing, T. (1990, November 8). Away with the dolphins. *The Independent Monthly*. https://wwat.nz/wp-content/uploads/1990-3.pdf.
3. Ozhiganova, A. (2022, January 24). Giving birth to a baby dolphin: Esoteric representations of human-dolphin connections in the late Soviet waterbirth movement. *Baltic Worlds*. https://balticworlds.com/giving-birth-to-a-baby-dolphin/.
4. Ozhiganova, A. (2022, January 24). Giving birth to a baby dolphin: Esoteric representations of human-dolphin connections in the late Soviet waterbirth movement. *Baltic Worlds*. https://balticworlds.com/giving-birth-to-a-baby-dolphin/.
5. King, D. (1990. September 23). Every mum wants her baby's birth to be special. But few would go as far as Sarah Evans, whose dream was to give birth with dolphins as midwives! *Woman's Own*. https://www.gentlebirthmethod.com/press/press-cuttings-dolphins/i-gave-birth-with-dolphins.
6. Ozhiganova, A. (2022, January 24). Giving birth to a baby dolphin: Esoteric representations of human-dolphin connections in the late Soviet waterbirth movement. *Baltic Worlds*. https://balticworlds.com/giving-birth-to-a-baby-dolphin/.
7. Ozhiganova, A. (2022, January 24). Giving birth to a baby dolphin: Esoteric representations of human-dolphin connections in the late Soviet waterbirth movement. *Baltic Worlds*. https://balticworlds.com/giving-birth-to-a-baby-dolphin/.
8. Tonetti, E. (1995). Waterbirth. Birth into Being. https://www.birthintobeing.com/waterbirth.
9. Caney, M (2013, May 28). Dolphin assisted birth: Therapeutic or wishful thinking? Dolphin Way. https://www.dolphin-way.com/2013/05/dolphin-assisted-birth-therapeutic-or-wishful-thinking/.
10. Dolphins and DNA (n.d.). http://www.dolphinspiritofhawaii.com/dolphins_and_dna.html.

NOTES

11. Stanley, A. (1995, June 8). A birth method stirs a debate. *New York Times.* https://www.nytimes.com/1995/06/08/garden/a-birth-method-stirs-a-debate.html.
12. Goswami, N. (2005, October 23). Dolphin sounds "aid brain growth in unborn child." *The Telegraph.* https://www.telegraph.co.uk/news/uknews/1501264/Dolphin-sounds-aid-brain-growth-in-unborn-child.html.
13. Goswami, N. (2005, October 23). Dolphin sounds "aid brain growth in unborn child." *The Telegraph.* https://www.telegraph.co.uk/news/uknews/1501264/Dolphin-sounds-aid-brain-growth-in-unborn-child.html.
14. Caney, M. (2013, May 28). Dolphin assisted birth: Therapeutic or wishful thinking? Dolphin Way. https://www.dolphin-way.com/2013/05/dolphin-assisted-birth-therapeutic-or-wishful-thinking/.
15. Chek, P. (host). (2019). Ep 27: Kim Nelli: Conscious birthing. *Living 4D with Paul Chek* [podcast]. https://podcasts.apple.com/fi/podcast/ep-27-kim-nelli/id1447548149?i=1000436893283.
16. *Katie Piper's Extraordinary Births* [film]. (2015). Channel 4.
17. Romanes, G. J. (1873). Permanent variation of colour in fish. *Nature, 8*(188), 101.
18. Darwin, C. (1872). *The expression of the emotions in man and animals* (3rd ed.). Ed. P. Ekman. Oxford University Press.
19. Darwin, C. *The expression of the emotions in man and animals.*
20. Romanes, G. J. (1882). *Animal Intelligence: The International Scientific Series* (Vol. 1). Library of Alexandria.
21. Romanes, G. J. *Animal Intelligence: The International Scientific Series.*
22. Thomson, A. (1879). Suicide of the scorpion. *Nature, 20*(520), 577.
23. Hutchinson, H. F. (1879). The bis-cobra, the goh-sámp, and the scorpion. *Nature, 20* (519), 553.
24. Morgan, C. L. (1883). Scorpion suicide. Nature, 27(701), 530–530.
25. Morgan, C. L. (1894). *An introduction to comparative psychology.* Walter Scott.
26. Morgan, C. L. (1885). *The springs of conduct: An essay on evolution.* Kegan Paul, Trench & Co.
27. Morgan, C. L. *An introduction to comparative psychology.*
28. Neave, N., McCarty, K., Freynik, J., Caplan, N., Hönekopp, J., & Fink, B. (2011). Male dance moves that catch a woman's eye. *Biology Letters, 7*(2), 221–224.
29. Schein, M. W., & Hale, E. B. (1957). The head as a sexual stimulus for orientation and arousal of sexual behavior of male turkeys. *Anatomical Record, 128,* 617–618.
30. A great review of Morgan's ideas can be found in this article: Fitzpatrick, S., & Goodrich, G. (2017). Building a science of animal minds: Lloyd Morgan, experimentation, and Morgan's canon. *Journal of the History of Biology, 50,* 525–569.

31. Thorndike, E. (2017). *Animal intelligence: Experimental studies*. Routledge.
32. Skinner, B. F. (2016). Why I am not a cognitive psychologist. In *Approaches to Cognition* (pp. 79–90). Routledge.
33. Brigandt, I. (2005). The instinct concept of the early Konrad Lorenz. *Journal of the History of Biology, 38*, 571–608.
34. Griffin, D. R. (1976). *The question of animal awareness: Evolutionary continuity of mental experience*. Rockefeller University Press.
35. Griffin, D. R. *The question of animal awareness: Evolutionary continuity of mental experience*.
36. Bekoff, M. (1995). Cognitive ethology: The comparative study of animal minds. In W. Bechtel & G. Graham (Eds.), *Blackwell companion to cognitive science*. Blackwell.
37. Griffin, D. R. (1998). From cognition to consciousness. *Animal Cognition, 1*, 3–16.
38. de Waal, F. B. (1999). Anthropomorphism and anthropodenial: Consistency in our thinking about humans and other animals. *Philosophical topics, 27*(1), 255–280.
39. Burghardt, G. M. (1985). Animal awareness: Current perceptions and historical perspective. *American Psychologist, 40*(8), 905.
40. Burghardt, G. M. Animal awareness: Current perceptions and historical perspective.
41. McKie, R. (2010, June 27). Chimps with everything: Jane Goodall's 50 years in the jungle. *The Guardian*. https://www.theguardian.com/science/2010/jun/27/jane-goodall-chimps-africa-interview.
42. High Profiles. (n.d.). Jane Goodall: High Profiles interview. https://highprofiles.info/interview/jane-goodall/.
43. Braitman, L. (2015). *Animal madness: inside their minds*. Simon and Schuster.
44. Safina, C. (2015, October). What are animals thinking and feeling? [Transcript]. TED.com. https://www.ted.com/talks/carl_safina_what_are_animals_thinking_and_feeling/transcript.
45. Wynne, C. D. (2004). The perils of anthropomorphism. *Nature, 428*(6983), 606.
46. Wynne, C. D. (2007). What are animals? Why anthropomorphism is still not a scientific approach to behavior. *Comparative Cognition & Behavior Reviews, 2*.
47. Bekoff, M. (2004). The great divide. *American Scientist, 92*(5), 481–483.
48. Bekoff, M. (2004). The great divide. *American Scientist, 92*(5), 481–483.
49. de Waal, F. B. (1991). Complementary methods and convergent evidence in the study of primate social cognition. *Behaviour, 118*(3–4), 297–320.
50. Wynne, C. D. L. (2005). The emperor's new anthropomorphism. *Behavior Analyst Today, 6*(3), 151–154.

NOTES

51. Kwai, I. (2023, November 15). "Everybody is a bit on edge": Sailors trade tips on steering clear of orcas. *New York Times.* https://www.nytimes.com/2023/11/15/world/europe/orcas-attacks-boats.html.
52. Carty, M. (2023, July 2). Orcas are ramming into boats, but experts warn against calling it revenge on humans. CBC. https://www.cbc.ca/radio/thecurrent/orca-boat-incidents-social-media-reaction-1.6895465.
53. Neuman, S. (2024, May 15). Orcas sank a yacht off Spain—the latest in a slew of such "attacks" in recent years. NPR. https://www.npr.org/2023/06/13/1181693759/orcas-killer-whales-boat-attacks.
54. Open Letter regarding Iberian orcas and their interactions with boats. (2023, August). Whale and Dolphin Conservation. https://uk.whales.org/wp-content/uploads/sites/6/2024/04/Open-letter-re-Iberian-orcas-FINAL-Sept2023.pdf
55. Zerbini, A. N., et al. (2024). *Workshop: Interactions between Iberian killer whales and vessels: Management recommendations.* Madrid, Spain, February 6–8, 2024. International Whaling Commission. https://archive.iwc.int/pages/download.php?direct=1&noattach=true&ref=22172&ext=pdf&k=.
56. Buckiewicz, A. (2024, May 31). Killer whales are ramming boats for fun, scientists say. A new report offers ways to protect mariners. *Quirks and Quarks.* CBC. https://www.cbc.ca/radio/quirks/killer-whale-spain-play-1.7220869.
57. Wu, K. (2024, June 5). The new anthropomorphism. *The Atlantic.* https://www.theatlantic.com/science/archive/2024/06/new-anthropomorphism/678611/.

Chapter 3: Wally the Alligator
1. Dahl, R. (2007). *The twits.* Penguin.
2. Wally was stolen and presumably released into the wild. At the time of writing this chapter, he had not been found/recovered. See Treistman, R. (2024, May 3). Have you seen this emotional support gator? Wally's owner says he's lost in Georgia. NPR. https://www.npr.org/2024/05/03/1248880788/wally-alligator-missing-georgia.
3. Treisman, R. (2023, September 29). Wally the emotional support alligator went to see the Phillies. Then he went viral. NPR. https://www.npr.org/2023/09/29/1202615294/wally-emotional-support-alligator-phillies-game.
4. Florida's Wildest. (2022, April 27). The problematic "emotional support alligator"—our unfiltered thoughts. YouTube. [video] https://www.youtube.com/watch?v=YWjPj9UYaOI.
5. Burghardt, G. M. (2014). A brief glimpse at the long evolutionary history of play. *Animal Behavior and Cognition, 1*(2), 90–98.
6. Dinets, V. (2015). Play behavior in crocodilians. *Animal Behavior and Cognition, 2*(1), 49–55.

NOTES

7. Chavan, U. M., & Borkar, M. R. (2023). Observations on cooperative fishing, use of bait for hunting, propensity for marigold flowers and sentient behaviour in Mugger crocodiles Crocodylus palustris (Lesson, 1831) of river Savitri at Mahad, Maharashtra, India. *Journal of Threatened Taxa*, 15(8), 23750–23762.
8. Dinets, V. (2015). Apparent coordination and collaboration in cooperatively hunting crocodilians. *Ethology, Ecology & Evolution*, 27(2), 244–250.
9. Dinets, V., Brueggen, J. C., & Brueggen, J. D. (2015). Crocodilians use tools for hunting. *Ethology, Ecology & Evolution*, 27(1), 74–78.
10. Rádlová, S., Janovcová, M., Sedláčková, K., Polák, J., Nácar, D., Pelešková, Š.,…& Landová, E. (2019). Snakes represent emotionally salient stimuli that may evoke both fear and disgust. *Frontiers in Psychology*, 10, 1085.
11. Janovcová, M., Rádlová, S., Polák, J., Sedláčková, K., Pelešková, Š., Žampachová, B.,…& Landová, E. (2019). Human attitude toward reptiles: A relationship between fear, disgust, and aesthetic preferences. *Animals*, 9(5), 238.
12. Albert, C., Luque, G. M., & Courchamp, F. (2018). The twenty most charismatic species. *PloS One*, 13(7), e0199149.
13. Bekoff, M. (1995). Play signals as punctuation: The structure of social play in canids. *Behaviour*, 132(5–6), 419–429.
14. Lambert, H., Carder, G., & D'Cruze, N. (2019). Given the cold shoulder: A review of the scientific literature for evidence of reptile sentience. *Animals*, 9(10), 821.
15. Burghardt, G. M. (2015). Play in fishes, frogs and reptiles. *Current Biology*, 25(1), R9–R10.
16. The Dodo. (2020, August 11). Woman has removed over 300 hooks from sharks' mouths | The Dodo Wild Hearts. YouTube [video]. https://www.youtube.com/watch?v=G8LmxwOgBhA.
17. Thomas, P. (2018, March 20). Cristina Zenato's remarkable ability with sharks allows for a hands-on approach toward helping them. *USA Today*. https://ftw.usatoday.com/2018/03/her-remarkable-ability-with-sharks-allows-for-a-hands-on-approach-toward-helping-them.
18. The Dodo. (2020, March 1). Diver has been friends with tiger shark for 20 years! | The Dodo. YouTube [video]. https://www.youtube.com/watch?v=FYonjn1oYcQ.
19. Darwin, C. (1900). *The life and letters of Charles Darwin*. D. Appleton.
20. Sheehan, M. J., & Tibbetts, E. A. (2011). Specialized face learning is associated with individual recognition in paper wasps. *Science*, 334(6060), 1272–1275.
21. Tibbetts, E. A., Pardo-Sanchez, J., Ramirez-Matias, J., & Avarguès-Weber, A. (2021). Individual recognition is associated with holistic face processing in Polistes paper wasps in a species-specific way. *Proceedings of the Royal Society B*, 288(1943), 20203010.

22. Erickson, J. (2011, December 1). Like humans, the paper wasp has a special talent for learning faces. *University of Michigan News*. https://news.umich.edu/like-humans-the-paper-wasp-has-a-special-talent-for-learning-faces/.
23. Cohen, T. (2011, December 1). Wasps are as good as humans when it comes to recognising each other's faces. *Daily Mail*. https://www.dailymail.co.uk/sciencetech/article-2068773/Wasps-good-humans-comes-recognising-faces.html.
24. Diaz, J. (2011, December 1). Motherf*cking wasps recognize each other by their faces. *Gizmodo*. https://gizmodo.com/motherf-cking-wasps-recognize-each-other-by-their-faces-5864566.
25. Lambert, H., Carder, G., & D'Cruze, N. (2019). Given the cold shoulder: A review of the scientific literature for evidence of reptile sentience. *Animals*, 9(10), 821.
26. Trimble, M. J., & Van Aarde, R. J. (2010). Species inequality in scientific study. *Conservation Biology*, 24(3), 886–890.
27. Proctor, H. S., Carder, G., & Cornish, A. R. (2013). Searching for animal sentience: A systematic review of the scientific literature. *Animals*, 3(3), 882–906.
28. Andrews, K. (n.d.). What is it like to be a crab? *Aeon*. https://aeon.co/essays/are-we-ready-to-study-consciousness-in-crabs-and-the-like.
29. Crick, F., & Koch, C. (1990). Towards a neurobiological theory of consciousness. In *Seminars in the Neurosciences*, Vol. 2 (pp. 263–275). Saunders Scientific Publications.
30. Low, P. (2012). The Cambridge Declaration on Consciousness. Proceedings of the Francis Crick Memorial Conference, Churchill College, Cambridge University, July 7, 2012 (pp. 1–2). The New York Declaration on Animal Consciousness. (2024, April 19). New York University. https://sites.google.com/nyu.edu/nydeclaration/declaration.
31. Chittka, L. (2023, July 1). Do insects feel joy and pain? *Scientific American*. https://www.scientificamerican.com/article/do-insects-feel-joy-and-pain/.

Chapter 4: Stove Spiders

1. Choi, C. (2008, May 5). A bug's sex life: A Q&A with Isabella Rossellini. *Scientific American*. https://www.scientificamerican.com/article/green-porno/.
2. Rakison, D. H., & Derringer, J. (2008). Do infants possess an evolved spider-detection mechanism? *Cognition*, 107(1), 381–393.
3. Isbister, G. K., & White, J. (2004). Clinical consequences of spider bites: Recent advances in our understanding. *Toxicon*, 43(5), 477–492.
4. Ingaham, C. (2015, June 16). Chart: The animals that are most likely to kill you this summer. *Washington Post*. https://www.washingtonpost.com

/news/wonk/wp/2015/06/16/chart-the-animals-that-are-most-likely-to-kill-you-this-summer/.
5. Isbister, G. K., Gray, M. R., Balit, C. R., Raven, R. J., Stokes, B. J., Porges, K.,...& Fisher, M. M. (2005). Funnel-web spider bite: A systematic review of recorded clinical cases. *Medical journal of Australia, 182*(8), 407–411.
6. Australian Bureau of Statistics. (2022). Causes of death, Australia. https://www.abs.gov.au/statistics/health/causes-death/causes-death-australia/latest-release.
7. Epley, N., Waytz, A., & Cacioppo, J. T. (2007). On seeing human: A three-factor theory of anthropomorphism. *Psychological Review, 114*(4), 864.
8. Kahneman, D. (2011). *Thinking, fast and slow*. Macmillan.
9. Greene, J. D. (2014). Beyond point-and-shoot morality: Why cognitive (neuro) science matters for ethics. *Ethics, 124*(4), 695–726.
10. For a review of these studies of anthropomorphism in child development, see: Goldman, E. J., & Poulin-Dubois, D. (2023). Children's anthropomorphism of inanimate agents. *Wiley Interdisciplinary Reviews: Cognitive Science*, e1676.
11. Simion, F., & Giorgio, E. D. (2015). Face perception and processing in early infancy: Inborn predispositions and developmental changes. *Frontiers in Psychology, 6*, 969.
12. Aschersleben, G., Hofer, T., & Jovanovic, B. (2008). The link between infant attention to goal-directed action and later theory of mind abilities. *Developmental Science, 11*(6), 862–868.
13. White, R. C., & Remington, A. (2019). Object personification in autism: This paper will be very sad if you don't read it. *Autism, 23*(4), 1042–1045; Tahiroglu, D., & Taylor, M. (2019). Anthropomorphism, social understanding, and imaginary companions. *British Journal of Developmental Psychology, 37*(2), 284–299.
14. Kana, R. K., Maximo, J. O., Williams, D. L., Keller, T. A., Schipul, S. E., Cherkassky, V. L.,...& Just, M. A. (2015). Aberrant functioning of the theory-of-mind network in children and adolescents with autism. *Molecular Autism, 6*(1), 1–12.
15. Clutterbuck, R. A., Shah, P., Leung, H. S., Callan, M. J., Gjersoe, N., & Livingston, L. A. (2022). Anthropomorphic tendencies in autism: A conceptual replication and extension of White and Remington (2019) and preliminary development of a novel anthropomorphism measure. *Autism, 26*(4), 940–950.
16. Tahiroglu, D., & Taylor, M. (2019). Anthropomorphism, social understanding, and imaginary companions. *British Journal of Developmental Psychology, 37*(2), 284–299; Epley, N., Akalis, S., Waytz, A., & Cacioppo, J. T. (2008). Creating social connection through inferential reproduction: Loneliness and perceived agency in gadgets, gods, and greyhounds. *Psychological Science, 19*(2), 114–120.

NOTES

17. Lim, M. H., Manera, K. E., Owen, K. B., Phongsavan, P., & Smith, B. J. (2023). The prevalence of chronic and episodic loneliness and social isolation from a longitudinal survey. *Scientific Reports, 13*(1), 12453.
18. Epley, N., Akalis, S., Waytz, A., & Cacioppo, J. T. (2008). Creating social connection through inferential reproduction: Loneliness and perceived agency in gadgets, gods, and greyhounds. *Psychological Science, 19*(2), 114–120.
19. Manfredo, M. J., Urquiza-Haas, E. G., Carlos, A. W. D., Bruskotter, J. T., & Dietsch, A. M. (2020). How anthropomorphism is changing the social context of modern wildlife conservation. *Biological Conservation, 241*, 108297.
20. De Wied, M., Meeus, W., & Van Boxtel, A. (2021). Disruptive behavior disorders and psychopathic traits in adolescents: Empathy-related responses to witnessing animal distress. *Journal of Psychopathology and Behavioral Assessment, 43*(4), 869–881; American Psychiatric Association. (2013). *Diagnostic and statistical manual of mental disorders: DSM-5.* American Psychiatric Association.
21. Wegner, D. M., & Gray, K. (2017). *The mind club: Who thinks, what feels, and why it matters.* Penguin.
22. American Psychiatric Association. (2013). *Diagnostic and statistical manual of mental disorders: DSM-5.* American Psychiatric Association; Jakobwitz, S., & Egan, V. (2006). The dark triad and normal personality traits. *Personality and Individual Differences, 40*(2), 331–339.
23. Gagne, P. (2024). *Sociopath: A memoir.* Simon & Schuster.
24. Buckels, E. E., Jones, D. N., & Paulhus, D. L. (2013). Behavioral confirmation of everyday sadism. *Psychological Science, 24*(11), 2201–2209.
25. Kavanagh, P. S., Signal, T. D., & Taylor, N. (2013). The dark triad and animal cruelty: Dark personalities, dark attitudes, and dark behaviors. *Personality and Individual Differences, 55*(6), 666–670.
26. Dugnoille, J. (2021). *Dogs and cats in South Korea: Itinerant commodities.* Purdue University Press.
27. Blaznik, M. (2018). Training young killers: How butcher education might be damaging young people. *Journal of Animal Ethics, 8*(2), 199–215.
28. Eisnitz, G. A. (2006). *Slaughterhouse: The shocking story of greed, neglect, and inhumane treatment inside the U.S. meat industry.* Prometheus Books.
29. Neff, C. (2015). The Jaws effect: How movie narratives are used to influence policy responses to shark bites in Western Australia. *Australian Journal of Political Science, 50*(1), 114–127.
30. McGuire, D. (2023). How *Jaws* influence shark perception. Shark Stewards. https://sharkstewards.org/how-jaws-influenced-shark-perception/.
31. Laverne, L. (host). (2022, December 18). Steven Spielberg. *The Desert Island Discs* [podcast]. BBC. https://www.bbc.co.uk/sounds/play/m001g8m4.
32. Daniels, S. E., Fanelli, R. E., Gilbert, A., & Benson-Amram, S. (2019). Behavioral flexibility of a generalist carnivore. *Animal cognition, 22*, 387–396.

33. Waytz, A., Cacioppo, J., & Epley, N. (2010). Who sees human? The stability and importance of individual differences in anthropomorphism. *Perspectives on Psychological Science, 5*(3), 219–232.
34. Robinson, M. (2014). Animal personhood in Mi'kmaq perspective. *Societies, 4*(4), 672–688.
35. Robinson, M. Animal personhood in Mi'kmaq perspective.
36. Robinson, M. Animal personhood in Mi'kmaq perspective.
37. Hatano, G., Siegler, R. S., Richards, D. D., Inagaki, K., Stavy, R., & Wax, N. (1993). The development of biological knowledge: A multi-national study. *Cognitive Development, 8*(1), 47–62.

Chapter 5: Companion Cube

1. Weir, A. (2016). *The Martian: A novel*. Broadway Books.
2. Guthrie, S. E. (1997). Anthropomorphism: A definition and a theory. In R. W. Mitchell, N. S. Thompson, & H. L. Miles (Eds.), *Anthropomorphism, anecdotes, and animals* (pp. 50–58). State University of New York Press.
3. Epley, N. (2015). *Mindwise: Why we misunderstand what others think, believe, feel, and want*. Vintage.
4. Airenti, G. (2018). The development of anthropomorphism in interaction: Intersubjectivity, imagination, and theory of mind. *Frontiers in Psychology, 9*, 2136.
5. Freymann, S., & Elffers, J. (1999). *How are you peeling?: Foods with moods*. Scholastic Inc..
6. Csibra, G., Gergely, G., Bı́ró, S., Koos, O., & Brockbank, M. (1999). Goal attribution without agency cues: The perception of "pure reason" in infancy. *Cognition, 72*(3), 237–267.
7. Damiano, L., & Dumouchel, P. (2018). Anthropomorphism in human-robot co-evolution. *Frontiers in Psychology, 9*, 468.
8. Yang, S., Mok, B. K. J., Sirkin, D., Ive, H. P., Maheshwari, R., Fischer, K., & Ju, W. (2015, August). Experiences developing socially acceptable interactions for a robotic trash barrel. In *2015 24th IEEE International Symposium on Robot and Human Interactive Communication (RO-MAN)* (277–284). IEEE.
9. Heider, F., & Simmel, M. (1944). An experimental study of apparent behavior. *American Journal of Psychology, 57*(2), 243–259.
10. Morewedge, C. K., Preston, J., & Wegner, D. M. (2007). Timescale bias in the attribution of mind. *Journal of Personality and Social Psychology, 93*(1), 1.
11. Elliot, S. (2007, December 13). Beyond the box. *Games for Windows Magazine*.
12. Faylor, C. (2008, February 22). GDC 08: Portal creators on writing, multiplayer, government interrogation techniques. Shack News. http://www.shacknews.com/featuredarticle.x?id=784.
13. Elliot, S. (2007, December 13). Beyond the box. *Games for Windows Magazine*.

14. Arkenberg, M. (2017). Cuteness and control in Portal. In J. P. Dale (Ed.), *The Aesthetics and Affects of Cuteness* (56–74). Routledge.
15. Barry, E. (2019, April 21). Finland's hobbyhorse girls, once a secret society, now prance in public. *New York Times*. https://www.nytimes.com/2019/04/21/world/europe/finland-hobbyhorse-girls.html.
16. Airenti, G. (2018). The development of anthropomorphism in interaction: Intersubjectivity, imagination, and theory of mind. *Frontiers in Psychology, 9*, 2136.
17. Tahiroglu, D., & Taylor, M. (2019). Anthropomorphism, social understanding, and imaginary companions. *British Journal of Developmental Psychology, 37*(2), 284–299.
18. Severson, R. L., & Woodard, S. R. (2018). Imagining others' minds: The positive relation between children's role play and anthropomorphism. *Frontiers in Psychology, 9*, 2140.
19. Burke, C. L., & Copenhaver, J. G. (2004). Animals as people in children's literature. *Language Arts, 81*(3), 205–213.
20. Larsen, N. E., Lee, K., & Ganea, P. A. (2018). Do storybooks with anthropomorphized animal characters promote prosocial behaviors in young children? *Developmental Science, 21*(3), e12590.
21. Thorne, S. (2019, May 8). "And all who sail in her…." *Legion Magazine*. https://legionmagazine.com/and-all-who-sail-in-her/.
22. Armstrong, D. (2013, September 17). Emotional attachment to robots could affect outcome on battlefield. University of Washington News. https://www.washington.edu/news/2013/09/17/emotional-attachment-to-robots-could-affect-outcome-on-battlefield/.
23. Carpenter, J. (2013). The quiet professional: An investigation of US military explosive ordnance disposal personnel interactions with everyday field robots (doctoral dissertation, University of Washington).
24. Marsh, A. (2010). Love among the objectum sexuals. *Electronic Journal of Human Sexuality, 13*.
25. Hugo, V. (1993). *The Hunchback of Notre-Dame*. Wordsworth Editions.
26. Scott, E. (2020, December 29). Woman marries her briefcase after five-year relationship. *Metro*. https://metro.co.uk/2020/12/29/woman-marries-her-briefcase-after-five-year-relationship-13818801/amp/.
27. Marsh, A. (2010). Love among the objectum sexuals. *Electronic Journal of Human Sexuality, 13*.

Chapter 6: Creepy Counterfeits
1. Shelley, M. W. (1998). *Frankenstein, or, The modern Prometheus: The 1818 text*. Oxford University Press.
2. Iley, C. (2012, February 5). So Jason Segel, are you a man or a Muppet? *The Guardian*. https://www.theguardian.com/culture/2012/feb/05/jason-segel-muppets-film-comedy.
3. Iley, C. (2012, February 5). So Jason Segel, are you a man or a Muppet?

The Guardian. https://www.theguardian.com/culture/2012/feb/05/jason-segel-muppets-film-comedy.

4. Austin Film Festival. (2017, May 27). On Story 707: Freaks & Muppets: A conversation with Jason Segel. YouTube [video] https://www.youtube.com/watch?v=tQWk7nZMSR4.
5. Austin Film Festival. (2017, May 27). On Story 707: Freaks & Muppets: A conversation with Jason Segel. YouTube [video] https://www.youtube.com/watch?v=tQWk7nZMSR4.
6. Weintraub, S. (2011, June 20). Jason Segel on set interview THE MUPPETS. *Collider.* https://collider.com/jason-segel-interview-the-muppets/97339/.
7. Why puppets? (2021, August 25). Puppet Nerd. https://puppetnerd.com/why-puppets/.
8. Rawlings, K. (2004). The earliest archaeological and textual evidence of puppets and shadows. UNESCO-APPAN Symposium and Festival. http://puppetryhistory.com/index.php/the-earliest-archaeological-and-textual-evidence-of-puppets-and-shadows/.
9. Bender, L., & Woltmann, A. G. (1936). The use of puppet shows as a psychotherapeutic method for behavior problems in children. *American Journal of Orthopsychiatry, 6*(3), 341–354.
10. Macari, S., Chen, X., Brunissen, L., Yhang, E., Brennan-Wydra, E., Vernetti, A.,...& Chawarska, K. (2021). Puppets facilitate attention to social cues in children with ASD. *Autism Research, 14*(9), 1975–1985.
11. Cummings, M. (2021, August 5). Study finds children with autism respond well to puppets. *Yale News.* https://news.yale.edu/2021/08/05/study-finds-children-autism-respond-well-puppets.
12. Robertson, J. (2020). No place for robots: Reassessing the bukimi no tani ("uncanny valley"). *Asia-Pacific Journal / Japan Focus, 18*(23, 4), 5521.
13. Kageki, N. (2012, June 12). An uncanny mind: Masahiro Mori on the uncanny valley and beyond. *IEEE Spectrum.* https://spectrum.ieee.org/an-uncanny-mind-masahiro-mori-on-the-uncanny-valley.
14. Clinton, P. (2004, November 10). Review: "Polar Express" a creepy ride. CNN. https://edition.cnn.com/2004/SHOWBIZ/Movies/11/10/review.polar.express/.
15. Smith, D. L. (n.d.). A theory of creepiness. *Aeon.* https://aeon.co/essays/what-makes-clowns-vampires-and-severed-hands-creepy.
16. For a summary of these hypotheses, read: Wang, S., Lilienfeld, S. O., & Rochat, P. (2015). The uncanny valley: Existence and explanations. *Review of General Psychology, 19*(4), 393–407.
17. Burleigh, T. J., Schoenherr, J. R., & Lacroix, G. L. (2013). Does the uncanny valley exist? An empirical test of the relationship between eeriness and the human likeness of digitally created faces. *Computers in Human Behavior, 29*(3), 759–771.
18. Kjeldgaard-Christiansen, J., & Clasen, M. (2023). Creepiness and the uncanny. *Style, 57*(3), 322–349.

NOTES

19. Gray, K., & Wegner, D. M. (2012). Feeling robots and human zombies: Mind perception and the uncanny valley. *Cognition, 125*(1), 125–130.
20. Wegner, D. M., & Gray, K. (2017). *The mind club: Who thinks, what feels, and why it matters*. Penguin.
21. Lischetzke, T., Izydorczyk, D., Hüller, C., & Appel, M. (2017). The topography of the uncanny valley and individuals' need for structure: A nonlinear mixed effects analysis. *Journal of Research in Personality, 68*, 96–113.
22. Castelo, N., & Sarvary, M. (2022). Cross-cultural differences in comfort with humanlike robots. *International Journal of Social Robotics, 14*(8), 1865–1873.
23. Feng, S., Wang, X., Wang, Q., Fang, J., Wu, Y., Yi, L., & Wei, K. (2018). The uncanny valley effect in typically developing children and its absence in children with autism spectrum disorders. *PloS One, 13*(11), e0206343.
24. Tinwell, A. (2015). Is the uncanny valley a universal or individual response? *Interaction Studies, 16*(2), 180–185.
25. Yam, K. C., Bigman, Y., & Gray, K. (2021). Reducing the uncanny valley by dehumanizing humanoid robots. *Computers in Human Behavior, 125*, 106945.
26. Park, G. (2019, November 12). We're surprised too: Sonic the Hedgehog's new look actually works. Here's why. *Washington Post*. https://www.washingtonpost.com/video-games/2019/11/12/were-surprised-too-sonic-hedgehogs-new-look-actually-works-heres-why/.
27. Russell, B. (2020, February 12). The Sonic movie director talks redesign: "It was pretty clear on the day the trailer was released that fans were not happy." Games Radar. https://www.gamesradar.com/sonic-movie-redesign-jeff-fowler-director-interview/.
28. Tyson, P. J., Davies, S. K., Scorey, S., & Greville, W. J. (2023). Fear of clowns: An investigation into the aetiology of coulrophobia. *Frontiers in Psychology, 14*, 1109466.
29. Echbiri, M. (2023, January 5). "M3GAN" producer James Wan shares how "Poltergeist" kindled his love for creepy dolls. *Collider*. https://collider.com/megan-movie-james-wan-creepy-doll-comments/.
30. Smith, D. L. (n.d.). A theory of creepiness. *Aeon*. https://aeon.co/essays/what-makes-clowns-vampires-and-severed-hands-creepy.
31. Tyson, P. J., Davies, S. K., Scorey, S., & Greville, W. J. (2023). Fear of clowns: An investigation into the aetiology of coulrophobia. *Frontiers in Psychology, 14*, 1109466.
32. 20-somethings are paying up for cosmetic procedures to prevent aging. (2024, January 16). CBS News. https://www.cbsnews.com/miami/news/20-somethings-are-paying-up-for-cosmetic-procedures-to-prevent-aging/.
33. Tyson, P. J., Davies, S. K., Scorey, S., & Greville, W. J. (2023). Fear of clowns: An investigation into the aetiology of coulrophobia. *Frontiers in Psychology, 14*, 1109466.

NOTES

34. Botox impairs ability to understand emotions of others. (2011, April 22). *USC Today*. https://today.usc.edu/botox-impairs-ability-to-understand-emotions-of-others/.
35. Phillips, M. (2020, October 23). Horror masks are never just about the monster. *New York Times*. https://www.nytimes.com/2020/10/23/movies/halloween-horror-masks.html.
36. Nitschinsk, L., Tobin, S. J., Varley, D., & Vanman, E. J. (2023). Why do people sometimes wear an anonymous mask? Motivations for seeking anonymity online. *Personality and Social Psychology Bulletin*, 01461672231210465; Silke, A. (2003). Deindividuation, anonymity, and violence: Findings from Northern Ireland. *Journal of Social Psychology*, 143(4), 493–499.
37. Kjeldgaard-Christiansen, J., & Clasen, M. (2023). Creepiness and the uncanny. *Style*, 57(3), 322–349.
38. Fleming, K. (2023, November 23). Inside the Cabbage Patch Kids craze that turned Black Friday shopping into a rite of violence. *New York Post*. https://nypost.com/2023/11/23/lifestyle/inside-the-cabbage-patch-kids-craze-that-fueled-black-friday/.
39. Friedrich, O. (1983, December 12). The strange Cabbage Patch craze. *Time*. https://content.time.com/time/subscriber/article/0,33009,921419-03,00.html.
40. Taylor, S. (2008, July 17). Attract, repel: Lifelike dolls are collector cult. Reuters. https://www.reuters.com/article/idUSL03734403/.
41. Taylor, S. (2008, July 17). Attract, repel: Lifelike dolls are collector cult. Reuters. https://www.reuters.com/article/idUSL03734403/.
42. Kugler, L. (2022, August 1). Crossing the uncanny valley. *Communications of the ACM*. https://cacm.acm.org/news/crossing-the-uncanny-valley/.
43. Boston Dynamics. (2024, April 29). Meet Sparkles. YouTube. [video]. https://www.youtube.com/watch?v=MG4PPkCyJig.
44. Destephe, M., Brandao, M., Kishi, T., Zecca, M., Hashimoto, K., & Takanishi, A. (2015). Walking in the uncanny valley: Importance of the attractiveness on the acceptance of a robot as a working partner. *Frontiers in Psychology*, 6, 204.
45. Kugler, L. (2022, August 1). Crossing the uncanny valley. *Communications of the ACM*. https://cacm.acm.org/news/crossing-the-uncanny-valley/.

Chapter 7: AI Overlords
1. Garland, A. (Dir.). (2015). *Ex Machina* [film]. Universal Pictures.
2. Putrill, J. (2023, February 28). Replika users fell in love with their AI chatbot companions. Then they lost them. ABC News. https://www.abc.net.au/news/science/2023-03-01/replika-users-fell-in-love-with-their-ai-chatbot-companion/102028196.
3. Funnell, A. (host). (2023, May 13). Falling in love with an app! when Anthropomorphism, making things too human like, goes wrong. *Future Tense*

NOTES

[podcast]. ABC. https://www.abc.net.au/listen/programs/futuretense/falling-in-love-with-an-app-anthropomorphism/102262704.

4. Putrill, J. (2023, February 28). Replika users fell in love with their AI chatbot companions. Then they lost them. ABC News. https://www.abc.net.au/news/science/2023-03-01/replika-users-fell-in-love-with-their-ai-chatbot-companion/102028196.
5. Artificial intelligence: Italian SA clamps down on "Replika" chatbot: Too many risks to children and emotionally vulnerable individuals. (2023, February 3). Garante per la protezione dei data personali. https://www.garanteprivacy.it/home/docweb/-/docweb-display/docweb/9852506#english.
6. Putrill, J. (2023, February 28). Replika users fell in love with their AI chatbot companions. Then they lost them. ABC News. https://www.abc.net.au/news/science/2023-03-01/replika-users-fell-in-love-with-their-ai-chatbot-companion/102028196.
7. Funnell, A. (host). (2023, May 13). Falling in love with an app! When anthropomorphism, making things too human like, goes wrong. *Future Tense* [podcast]. ABC. https://www.abc.net.au/listen/programs/futuretense/falling-in-love-with-an-app-anthropomorphism/102262704.
8. Funnell, A. (host). (2023, May 13). Falling in love with an app! When anthropomorphism, making things too human like, goes wrong. *Future Tense* [podcast]. ABC. https://www.abc.net.au/listen/programs/futuretense/falling-in-love-with-an-app-anthropomorphism/102262704.
9. Weizenbaum, J. (1966). ELIZA—a computer program for the study of natural language communication between man and machine. *Communications of the ACM, 9*(1), 36–45.
10. Weizenbaum, J. (1976). *Computer power and human reason: From judgment to calculation*. W. H. Freeman.
11. Turing, A. M. (2009). Computing machinery and intelligence. In R. S. Epstein, G. Roberts, & G. Beber (Eds.), *Parsing the Turing test* (pp. 23–65). Springer Netherlands.
12. Bender, E. M., Gebru, T., McMillan-Major, A., & Shmitchell, S. (2021, March). On the dangers of stochastic parrots: Can language models be too big? In *Proceedings of the 2021 ACM conference on fairness, accountability, and transparency* (pp. 610–623). ACM.
13. Weizenbaum, J. (1976). *Computer power and human reason: From judgment to calculation*. W. H. Freeman.
14. Seth, A. (2021). *Being you: A new science of consciousness*. Penguin.
15. Klein, E. (host). (2024, April 2). How should I be using A.I. right now? *The Ezra Klein Show* [podcast]. https://www.nytimes.com/2024/04/02/opinion/ezra-klein-podcast-ethan-mollick.html.
16. Claude. AI assistant created by Anthropic. Accessed April 17, 2024.

NOTES

17. Chalmers, D. J. (2023). Could a large language model be conscious? arXiv preprint. arXiv:2303.07103.
18. Seth, A. (2023, May 8). Why conscious AI is a bad, bad idea. *Nautilus*. https://nautil.us/why-conscious-ai-is-a-bad-bad-idea-302937/.
19. Seth, A. (2023, May 8). Why conscious AI is a bad, bad idea. *Nautilus*. https://nautil.us/why-conscious-ai-is-a-bad-bad-idea-302937/.
20. Lemoine, B. (2022, June 11). Is LaMDA sentient? An interview. *Medium*. https://www.documentcloud.org/documents/22058315-is-lamda-sentient-an-interview.
21. From the marketing info found on the homepage of https://www.myyov.com/.
22. Reinboth, T. (2024, April 4). Do "griefbots" help mourners deal with loss? *Undark*. https://undark.org/2024/04/04/opinion-griefbots-lack-evidence/.
23. Wallach, W., & Allen, C. (2008). *Moral machines: Teaching robots right from wrong*. Oxford University Press.
24. Like https://www.sapan.ai/.
25. Gross, T. (host) (2006, September 18). Pastor John Hagee on Christian Zionism. *Fresh Air*. NPR. https://www.npr.org/2006/09/18/6097362/pastor-john-hagee-on-christian-zionism.
26. Gervais, W. M. (2013). Perceiving minds and gods: How mind perception enables, constrains, and is triggered by belief in gods. *Perspectives on Psychological Science*, 8(4), 380–394.
27. Nietzsche, F. W. (2006). *Thus spoke Zarathustra: A book for all and none*. Cambridge University Press.
28. Jackson, J. C., Dillion, D., Bastian, B., Watts, J., Buckner, W., DiMaggio, N., & Gray, K. (2023). Supernatural explanations across 114 societies are more common for natural than social phenomena. *Nature Human Behaviour*, 7(5), 707–717.
29. Lombrozo, T. (June 13, 2023) Why we hallucinate supernation explanations. *Nautilus*. https://nautil.us/why-we-believe-supernatural-explanations-310823/
30. Fitzpatrick, M. (2020, May 15). Quarantining with a ghost? It's scary. *New York Times*. https://www.nytimes.com/2020/05/14/style/haunted-house-ghost-quarantine.html.
31. Engle, J. (2020, May 20). Do you believe in ghosts? *New York Times*. https://www.nytimes.com/2020/05/20/learning/do-you-believe-in-ghosts.html.
32. Tropical cyclone naming history and retired names. (n.d.). National Hurricane Center and Central Pacific Hurricane Center. https://www.nhc.noaa.gov/aboutnames_history.shtml.
33. National Hurricane Center. (n.d.). History of naming hurricanes. National Oceanic and Atmospheric Administration. https://www.nhc.noaa.gov/aboutnames_history.shtml.
34. Ciciora, P. (2014, June 2). Study: Hurricanes with female names more

deadly than male-named storms. *University of Illinois News Bureau.* https://news.illinois.edu/view/6367/204580.

35. Robson, D. (2023, September 17). Why we personify threatening events. *Future.* BBC. https://www.bbc.com/future/article/20230914-why-we-personify-threatening-events.

36. Wang, L., Touré-Tillery, M., & McGill, A. L. (2023). The effect of disease anthropomorphism on compliance with health recommendations. *Journal of the Academy of Marketing Science, 51*(2), 266–285.

37. Allen, S. (2022, September 1). How humanizing disease could be a new public health tool. *Kellogg Insight.* https://insight.kellogg.northwestern.edu/article/how-humanizing-disease-could-be-a-new-public-health-tool.

38. Bullo, S., & Hearn, J. H. (2021). Parallel worlds and personified pain: A mixed-methods analysis of pain metaphor use by women with endometriosis. *British Journal of Health Psychology, 26*(2), 271–288.

39. Schattner, E., & Shahar, G. (2011). Role of pain personification in pain-related depression: An object relations perspective. *Psychiatry, 74*(1), 14–20; Tsur, N., Noyman-Veksler, G., Elbaz, I., Weisman, L., Brill, S., Shalev, H.,...& Shahar, G. (2023). The personification of chronic pain: An examination using the Ben-Gurion University Illness Personification Scale (BGU-IPS). *Psychiatry, 86*(2), 137–156; Aloni, R., Shahar, G., Ben-Ari, A., Margalit, D., & Achiron, A. (2023). Negative and positive personification of multiple sclerosis: Role in psychological adaptation. *Journal of Psychosomatic Research, 164,* 111078.

40. Hauser, D. J., & Schwarz, N. (2020). The war on prevention II: Battle metaphors undermine cancer treatment and prevention and do not increase vigilance. *Health Communication, 35*(13), 1698–1704.

41. Sample, I. (2019, August 10). "War on cancer" metaphors may do harm, research shows. *The Guardian.* https://www.theguardian.com/society/2019/aug/10/war-cancer-metaphors-harm-research-shows.

42. Skuse, A. (2014). Wombs, worms and wolves: Constructing cancer in early modern England. *Social History of Medicine, 27*(4), 632–648.

43. Guthrie, S., Agassi, J., Andriolo, K. R., Buchdahl, D., Earhart, H. B., Greenberg, M.,...& Tissot, G. (1980). A cognitive theory of religion [and comments and reply]. *Current Anthropology, 21*(2), 181–203; Guthrie, S. E. (1995). *Faces in the clouds: A new theory of religion.* Oxford University Press.

44. Guthrie, S., & Porubanova, M. (2020). Faces in clouds and voices in wind: Anthropomorphism in religion and human cognition. In *Routledge handbook of evolutionary approaches to religion.* Routledge.

45. Guthrie, S. (2013). Spiritual beings: A Darwinian, cognitive account. In G. Harvey (Ed.), *The handbook of contemporary animism* (353–358). Acumen Publishing.

46. Barrett, J. L. (2000). Exploring the natural foundations of religion. *Trends in Cognitive Sciences, 4*(1), 29–34.

NOTES

47. Gray, K., & Wegner, D. M. (2010). Blaming God for our pain: Human suffering and the divine mind. *Personality and Social Psychology Review, 14*(1), 7–16.
48. Wegner, D. M., & Gray, K. (2017). *The mind club: Who thinks, what feels, and why it matters*. Penguin.
49. Religion as anthropomorphism. (2014, February 10). *The Religious Studies Project* [podcast]. https://www.religiousstudiesproject.com/podcast/religion-as-anthropomorphism-an-interview-with-stewart-guthrie/.
50. Morley, J. (2023, October 28). Anthropic long-term benefit trust. Harvard Law School Forum on Corporate Governance. https://corpgov.law.harvard.edu/2023/10/28/anthropic-long-term-benefit-trust/.
51. Hu, K. (2023, October 27). Google agrees to invest up to $2 billion in OpenAI rival Anthropic. Reuters. https://www.reuters.com/technology/google-agrees-invest-up-2-bln-openai-rival-anthropic-wsj-2023-10-27/; Dastin, J. (2023, September 29). Amazon steps up AI race with Anthropic investment. Reuters. https://www.reuters.com/markets/deals/amazon-steps-up-ai-race-with-up-4-billion-deal-invest-anthropic-2023-09-25/; Schattner, E., & Shahar, G. (2011). Role of pain personification in pain-related depression: An object relations perspective. *Psychiatry, 74*(1), 14–20; Tsur, N., Noyman-Veksler, G., Elbaz, I., Weisman, L., Brill, S., Shalev, H.,…& Shahar, G. (2023). The personification of chronic pain: an examination using the Ben-Gurion University Illness Personification Scale (BGU-IPS). *Psychiatry, 86*(2), 137–156.
52. Bushwick, S. (host). (2023, April 28). Why do humans anthropomorphize AI? *Science Friday* [podcast]. https://www.sciencefriday.com/segments/ai-human-personification/#segment-transcript.

Chapter 8: Cute Capitalism

1. This quote is often attributed to marketing super guru Terry O'Reilly, who offers this as part of the marketing copy for his public speaking services: https://www.speakers.ca/speakers/terry-oreilly/.
2. Elliott, S. (September 16, 2002). The Media Business: advertising; Ikea challenges the attachment to old stuff, in favor of brighter, new stuff. *The New York Times*. https://www.nytimes.com/2002/09/16/business/media-business-advertising-ikea-challenges-attachment-old-stuff-favor-brighter.html
3. Elliott, S. The Media Business: advertising; Ikea challenges the attachment to old stuff, in favor of brighter, new stuff.
4. Lee, J. M., Baek, J., & Ju, D. Y. (2018). Anthropomorphic design: Emotional perception for deformable object. *Frontiers in Psychology, 9*, 387347.
5. See discussion in Yang, L. W., Aggarwal, P., & McGill, A. L. (2020). The 3 C's of anthropomorphism: Connection, comprehension, and competition. *Consumer Psychology Review, 3*(1), 3–19.
6. Schroll, R. (2023). "Ouch!" When and why food anthropomorphism

negatively affects consumption. *Journal of Consumer Psychology, 33*(3), 561–574.

7. Hartmann, J., Bergner, A., & Hildebrand, C. (2023). MindMiner: Uncovering linguistic markers of mind perception as a new lens to understand consumer–smart object relationships. *Journal of Consumer Psychology, 33*(4), 645–667.

8. Uysal, E., Alavi, S., & Bezençon, V. (2022). Trojan horse or useful helper? A relationship perspective on artificial intelligence assistants with humanlike features. *Journal of the Academy of Marketing Science, 50*(6), 1153–1175.

9. Yang, L. W., Aggarwal, P., & McGill, A. L. (2020). The 3 C's of anthropomorphism: Connection, comprehension, and competition. *Consumer Psychology Review, 3*(1), 3–19.

10. Fronczek, L. P., Mende, M., Scott, M. L., Nenkov, G. Y., & Gustafsson, A. (2023). Friend or foe? Can anthropomorphizing self-tracking devices backfire on marketers and consumers? *Journal of the Academy of Marketing Science, 51*(5), 1075–1097.

11. Sung, M. (2019, April 2). Why does the Duolingo owl scare me more than my high school Spanish teacher ever did? Mashable. https://mashable.com/article/duolingo-anxiety-notifications.

12. Rossiter, J., & Bellman, S. (2012). Emotional branding pays off: How brands meet share of requirements through bonding, companionship, and love. *Journal of Advertising Research, 52*(3), 291–296.

13. Blackett, W. (2022, October 13). Brand love and guilty pleasures: McDonald's shows we turn to comfort food in a crisis. *The Drum.* https://www.thedrum.com/opinion/2022/10/13/brand-love-and-guilty-pleasures-mcdonalds-shows-we-turn-comfort-food-crisis.

14. Rossiter, J., & Bellman, S. (2012). Emotional branding pays off: How brands meet share of requirements through bonding, companionship, and love. *Journal of Advertising Research, 52*(3), 291–296.

15. We are Panera Bread. (n.d.). Panera Bread. https://www.panera.ca/en-us/company/about-panera.html.

16. Cooke, L. (2018, April 6). The un-cuddly truth about pandas. *Wall Street Journal.* https://www.wsj.com/articles/the-un-cuddly-truth-about-pandas-1523025742.

17. Lorimer, J. (2007). Nonhuman charisma. *Environment and Planning D: Society and Space, 25*(5), 911–932.

18. McKirdy, E. (2014, May 12). Japanese cuteness overload could result in mascot cull. CNN. https://edition.cnn.com/2014/05/12/world/asia/osaka-mascot-cull/.

19. Fortuna, T. (2021, April 7). Ok, I've tried. I've tried for the last several months to post this dog for adoption and make him sound...palatable. Facebook post. https://www.facebook.com/tyfanee.fortuna/posts/10219752628710467.

20. Butterfield, M. E., Hill, S. E., & Lord, C. G. (2012). Mangy mutt or furry

friend? Anthropomorphism promotes animal welfare. *Journal of Experimental Social Psychology, 48*(4), 957–960.
21. Chan, A. A. Y. H. (2012). Anthropomorphism as a conservation tool. *Biodiversity and Conservation, 21,* 1889–1892.
22. Burnet, M. (2024, March 20). Rethinking anthropomorphism. Association of Zoos and Aquariums. https://www.aza.org/connect-stories/stories/rethinking-anthropomorphism.
23. Root-Bernstein, M., Douglas, L., Smith, A., & Verissimo, D. (2013). Anthropomorphized species as tools for conservation: Utility beyond prosocial, intelligent and suffering species. *Biodiversity and Conservation, 22,* 1577–1589.
24. McLeod, L. (2024, January 9). Should you keep a raccoon as a pet? The Spruce Pets. https://www.thesprucepets.com/pet-raccoons-1237219.
25. Ryall, J. (2019, September 25). Raccoons wreak havoc in Japan. Deutsche Welle. https://www.dw.com/en/top-stories/s-9097.
26. The Nobel Peace Prize for 2007. (October 12, 2007). Norwegian Nobel Committee. http://nobelpeaceprize.org/en_GB/laureates/laureates-2007/announce-2007/.
27. Speech by Al Gore on accepting the Nobel Peace Prize. (2007, December 11). Grist. https://grist.org/climate-energy/for-this-purpose-we-will-rise-and-we-will-act/.
28. Al Gore: There's still time to save the planet. (2006, June 23). ABC News. https://abcnews.go.com/GMA/GlobalWarming/story?id=2110628&page=1; Seabrook, A. (2007, March 21). Gore takes global warming message to Congress. NPR. https://www.npr.org/2007/03/21/9047642/gore-takes-global-warming-message-to-congress.
29. Khazan, O. (2013, November 7). The psychology of giant princess eyes. *The Atlantic.* https://www.theatlantic.com/health/archive/2013/11/the-psychology-of-giant-princess-eyes/281209/.
30. Schwartzel, E. (2018, October 24). Disney World's big secret: It's a favorite spot to scatter family ashes. *Wall Street Journal.* https://www.wsj.com/articles/disney-worlds-big-secret-its-a-favorite-spot-to-scatter-family-ashes-1540390229.
31. Brédart, S. (2021). The influence of anthropomorphism on giving personal names to objects. *Advances in Cognitive Psychology, 17*(1).
32. Keaveney, S. M., Herrmann, A., Befurt, R., & Landwehr, J. R. (2012). The eyes have it: How a car's face influences consumer categorization and evaluation of product line extensions. *Psychology & Marketing, 29*(1), 36–51.
33. Hoback, A. S. (2018). Pareidolia and perception of anger in vehicle styles: Survey results. *International Journal of Psychological and Behavioral Sciences, 12*(8), 1049–1055.
34. Aggarwal, P., & McGill, A. L. (2007). Is that car smiling at me? Schema congruity as a basis for evaluating anthropomorphized products. *Journal of Consumer Research, 34*(4), 468–479.

35. Ariella, S. (2023, March 15). 36 automotive industry statistics [2023]: Average employment, sales, and more. Zippa. https://www.zippia.com/advice/automotive-industry-statistics/; North America automotive market size & share analysis—growth trends & forecasts (2024–2029). (n.d.). Mordor Intelligence. https://www.mordorintelligence.com/industry-reports/north-america-automotive-market.
36. Miesler, L., Leder, H., & Herrmann, A. (2011). Isn't it cute: An evolutionary perspective of baby-schema effects in visual product designs. *International Journal of Design, 5*(3).
37. Rose, J. (2023, November 14). Taller cars and trucks are more dangerous for pedestrians, according to crash data. NPR. https://www.npr.org/2023/11/14/1212737005/cars-trucks-pedestrian-deaths-increase-crash-data.
38. Pazhoohi, F., & Kingstone, A. (2022). Larger vehicles are perceived as more aggressive, angry, dominant, and masculine. *Current Psychology, 41,* 4195–4199.
39. Zipper, D. (2020, August 12). The life-saving car technology no one wants. Bloomberg. https://www.bloomberg.com/news/features/2020-08-12/why-are-cars-still-so-dangerous-to-pedestrians.
40. Tyndall, J. (2024). The effect of front-end vehicle height on pedestrian death risk. *Economics of Transportation, 37,* 100342.
41. Young, J. (2021, February 11). Vehicle choice, crash differences help explain greater injury risks for women. Insurance Institute for Highway Safety. https://www.iihs.org/news/detail/vehicle-choice-crash-differences-help-explain-greater-injury-risks-for-women.
42. Torchinsky, J. (2018, December 4). We need to talk about truck design right now before it's too late. Jalopnik. https://jalopnik.com/we-need-to-talk-about-truck-design-right-now-before-its-1830860270.
43. Daggett, C. (2018). Petro-masculinity: Fossil fuels and authoritarian desire. *Millennium, 47*(1), 25–44. See also Schmitt, A. (2021, March 11). What happened to pickup trucks? Bloomberg. https://www.bloomberg.com/news/articles/2021-03-11/the-dangerous-rise-of-the-supersized-pickup-truck.
44. Albert, D. (n.d.). The American pickup. *n+1.* https://www.nplusonemag.com/online-only/online-only/the-american-pickup/.

Chapter 9: Puppy Propaganda
1. Huxley, A. (1936). *The olive tree and other essays.* Chatto & Windus.
2. Gee, A. (2014, January 6). Pushinka: A Cold War puppy the Kennedys loved. BBC. https://www.bbc.com/news/magazine-24837199.
3. This story is described in: Kamenetsky, C. (2019). *Children's literature in Hitler's Germany: The cultural policy of National Socialism.* Ohio University Press.
4. Hsu, C. (2015, August 28). How media "fluff" helped Hitler rise to

NOTES

power. University of Buffalo News Center. https://www.buffalo.edu/news/releases/2015/08/034.html.

5. Robinson, N. (2021, January 15). Puppaganda: How politicians use pets to convince you of their humanity. *Current Affairs*. https://www.currentaffairs.org/news/2021/01/dogaganda-how-politicians-use-pets-to-convince-you-of-their-humanity.
6. Dearden, L. (2016, August 2). ISIS using kittens and honey bees in bid to soften image in Dabiq propaganda magazine. *The Independent*. https://www.independent.co.uk/news/world/middle-east/isis-kittens-honey-bees-dabiq-propaganda-recruits-photo-soften-image-terror-a7168586.html.
7. Reynolds, L. (2015, December 21). ISIS recruiter uses pet cat to entice youngsters to join terror group. *Express*. https://www.express.co.uk/news/world/628560/Islamic-State-recruiter-uses-pet-cat-entice-youngsters-terror-group.
8. Zou, R., Lu, L., Cai, J., & Ran, Y. (2024). Does a cute pet make a difference in home-sharing booking intentions? A moderated serial mediation analysis. *International Journal of Hospitality Management, 118*, 103666.
9. Landwehr, J. R., & Wänke, M. (2023). Face-to-face: Three facial features that may turn the scale in close electoral races. *Journal of Experimental Social Psychology, 108*, 104488.
10. Llach, L. (2024, May 4). Meet Aitana, Spain's first AI model, who is earning up to €10,000 a month. *Euronews*. https://www.euronews.com/next/2024/05/04/meet-the-first-spanish-ai-model-earning-up-to-10000-per-month.
11. Waytz, A., Epley, N., & Cacioppo, J. T. (2010). Social cognition unbound: Insights into anthropomorphism and dehumanization. *Current Directions in Psychological Science, 19*(1), 58–62.
12. Wegner, D. M., & Gray, K. (2017). *The mind club: Who thinks, what feels, and why it matters*. Penguin.
13. Smith, D. L. (2021). *Making monsters: The uncanny power of dehumanization*. Harvard University Press.
14. Smith, D. L. *Making monsters: The uncanny power of dehumanization*.
15. Grossman, D. (2014). *On killing: The psychological cost of learning to kill in war and society*. Open Road Media.
16. The Prosecutor v. Jean-Paul Akayesu (Trial Judgement), ICTR-96-4-T, International Criminal Tribunal for Rwanda (ICTR), September 2, 1998, https://www.refworld.org/jurisprudence/caselaw/ictr/1998/en/19275 [accessed June 27, 2024].
17. Butler, J., Burns, D. P., & Robson, C. (2021). Dodgeball: Inadvertently teaching oppression in physical and health education. *European Physical Education Review, 27*(1), 27–40.
18. Leah, S. (n.d.). #Ableism. Center for Disability Rights. https://cdrnys.org/blog/uncategorized/ableism/.

NOTES

19. Sitruk, J., Summers, K. M., & Lloyd, E. P. (2023). Dehumanizing disability: Evidence for subtle and blatant dehumanization of people with physical disabilities. *Current Research in Ecological and Social Psychology, 5,* 100162.
20. Robertson, J. (2020, December 1). No place for robots: Reassessing the bukimi no tani ("uncanny valley"). *Asia-Pacific Journal: Japan Focus.* https://apjjf.org/2020/23/robertson.
21. Hoffman, K. M., Trawalter, S., Axt, J. R., & Oliver, M. N. (2016). Racial bias in pain assessment and treatment recommendations, and false beliefs about biological differences between blacks and whites. *Proceedings of the National Academy of Sciences, 113*(16), 4296–4301.
22. Todd, K. H., Deaton, C., D'Adamo, A. P., & Goe, L. (2000). Ethnicity and analgesic practice. *Annals of Emergency Medicine, 35*(1), 11–16.
23. Hoffman, K. M., Trawalter, S., Axt, J. R., & Oliver, M. N. (2016). Racial bias in pain assessment and treatment recommendations, and false beliefs about biological differences between blacks and whites. *Proceedings of the National Academy of Sciences, 113*(16), 4296–4301.
24. Fu, F., Tarnita, C. E., Christakis, N. A., Wang, L., Rand, D. G., & Nowak, M. A. (2012). Evolution of in-group favoritism. *Scientific Reports, 2*(1), 460.
25. Everett, J. A., Faber, N. S., & Crockett, M. (2015). Preferences and beliefs in ingroup favoritism. *Frontiers in Behavioral Neuroscience, 9,* 126656.
26. Waytz, A. (2019). *The power of human: How our shared humanity can help us create a better world.* W. W. Norton & Company.
27. Waytz, A., Epley, N., & Cacioppo, J. T. (2010). Social cognition unbound: Insights into anthropomorphism and dehumanization. *Current Directions in Psychological Science, 19*(1), 58–62.
28. Lincoln, A. (1953). *Collected Works of Abraham Lincoln, Vol. 1.* Rutgers University Press.

INDEX

A
Aarniomäki, Alisa, 129
Abernethy, Jim, 80
ableism, 12, 235–236
Adams, John Quincy, 228
Aeon (magazine), 84–86
Aesthetic Surgery Journal, 159
agency, 7–8
   AI and, 209
   dehumanization and, 231–235
   diseases and, 192–193
   effectance motivation hypothesis on, 10–11
   eyes, eyebrows, and, 28–29
   face recognition and, 93–96
   movement and, 119–121
   religion and, 194–195
AI. *See* artificial intelligence (AI)
AIAs. *See* artificial intelligence assistants (AIAs)
Airbnb, 229
Airenti, Gabriella, 117, 129
Aitana, 230
Akayesu, Jean-Paul, 234
Akeakamai the dolphin, 33
Alavi, Sascha, 208
Albert, Dan, 224
Alexa, 198
Alex the parrot, 33
Allen, Colin, 183
ambiguity, uncanny valley and, 151–154, 157–165, 230–231
Andrews, Kristin, 84–86
Animal Behavior and Cognition Lab, 107–108
Animal Intelligence (Romanes), 50–52
Animal Intelligence (Thorndike), 55–56
Animal Madness: Inside Their Minds (Braitman), 60
animals, 17–41
   anthropomorphism triggers in, 13–14
   babies, caregiver response, and, 20–25
   behaviorism on, 55–57
   breeding for cute behavior, 27–29
   breeding for cuteness, 25–230
   consciousness and cognition in, 29–39, 41, 43–66, 84–86, 107–108, 247–248
   crocodilians, 67–74, 228
   dolphins, 33, 43–49, 64–65
   environmental conservation and, 215–218
   ethology on, 57–59
   as false mirrors, 35–39
   fauxnads for neutered, 1–3, 15
   feelings of for humans, 29–30, 67–74, 79–80
   food culture and, 103–105
   growing up with, 97
   harming, empathy and, 108–112
   infant-directed speech and, 33–34
   intelligent behavior in, 74
   intra-species recognition and, 6
   language and, 171–172
   Lincoln's love of, 17–19, 21, 25, 228, 240–241
   mammalcentrism and, 83–87
   as mascots, 211–215
   orcas, 62–64
   owners eaten by, 39–40
   personality in, 71
   playfulness in, 76–79
   political use of, 225–229
   portrayal of in film and literature, 133–135
   pros and cons of anthropomorphizing, 247–249
   raccoons, 107–108, 216–217
   sadism toward, 103
   sharks, 79–80
   sounds made by, 24–25
   turkey mating behavior, 53–55
   ugly, 66–87
   utility of anthropomorphism in science and, 44–62
animism, 112–113, 155
Anthropic, 177–180, 197–198
anthropocentrism, 83–87
anthropodenial, 59
Anthropo-Dial, 125–135
   AI and, 172–173
   categorization failures and, 151–152
   conscious control over, 239–240
   definition of, 125
   dehumanization and, 231–235, 250

INDEX

Anthropo-Dial *(cont.)*
    downside of object anthropomorphization, 139–142
    fictional characters and, 131–135
    in play, 127–130
    uncanny valley and, 149–150
anthropomorphism
    benefits and dangers of, 244–251
    conscious control over, 14, 126–127, 239–240, 243–251
    criticisms of, 8–9
    definitions of, 4–5
    evolutionary benefits of, 9–12, 113
    fun in using, 15–16, 113, 124, 141–142, 247, 251
    how it works, 6–8
    individual and cultural differences in, 87, 89–113
    over-attribution and under-attribution in, 86–87
    in science, 4–5
    triggers of, 12–15
antisocial personality disorder, 102–103
Are We There Yet? The American Automobile Past, Present and Driverless (Albert), 224
Arkenberg, Megan, 124
art, 14
artificial intelligence (AI), 4–5, 14–15, 167, 169–199
    avoiding anthropomorphization of, 249–250
    capitalism and, 197–199
    chatbots, 169–171
    consciousness of, 176–182
    context incorporation by, 178–179
    delusional thinking about, 175–177
    griefbots, 182–184
    human distinctiveness threatened by, 208–209
    as influencers, 230–231
    marketing and, 207–209
    as minds, 171–176
    rights groups for, 183–184
artificial intelligence assistants (AIAs), 207–208
Atlas robot, 165
auditory cuteness, 24–25
Australian bush turkey, 22
autism, 95–96, 146–147, 155
autonomy, 209
axolotl, 22

B
babies
    caregiver response to, 20–25
    dolphin-assisted births and, 43–49
    facial cuteness of, 21–24, 27, 28
    infant-directed speech and, 33–34
    movement, intention, and, 119–121
    pain experienced by, 37–38
    Soviet superbabies, 44–47
Bacon, Francis, 8, 36
Barényi, Béla, 220
Barrett, Justin, 194
bat echolocation, 57–58
BBC, 190
Bed Bath & Beyond, 201, 202
bees, 57, 86
behaviorism, 55–57
Being You (Seth), 176–177
Bekoff, Marc, 4, 61
Bell, Kristen, 23
Bellman, Steven, 210
Benchley, Peter, 107
Bender, Emily, 175
Bezençon, Valéry, 208
biases
    cognitive, facial recognition and, 81–83
    in-group, 237–240
    motion and timescale, 121
Biology and Conservation (journal), 215–216
Blindsight (Watts), 171–175
boats, 136–137
Bogusky, Alex, 203
Boldrin, Sergio, 161
bomb disposal robots, 138–139
Boston Dynamics, 165, 166
Botox, 158–160
brachycephalic animals, 26–27
brain. *See* neurobiology
Braitman, Laurel, 60
breast cancer, 190–191
Bunny the dog, 30–31, 34
Burghardt, Gordon, 59, 72, 78
Burke, Carolyn, 131
Burleigh, Tyler, 152
Burnet, Marta, 216
butterflies, 99–101
Byosiere, Sarah-Elizabeth, 31

C
Cabbage Patch Kids, 163–164
Cacioppo, John, 5, 97–98, 109, 231, 238
Cambridge Declaration on Consciousness, 85
cancer, 190–191, 192–193
Canine Science Collaboratory, 31–32
Canine Testicular Implantation (CTI), 2–3
capitalism, 14–15, 197–199, 201–224
    Disney and, 219–220
    emotional branding and, 210–211
    IKEA and, 201–204
    mascots and, 211–215
caregiver response
    animals bred for cuteness and, 25–27
    babies, animals, and, 20–25
    eyes, eyebrows, and, 28–29
    fictional characters and, 135
    to inanimate objects, 121–124, 135–139
carriages, 221

INDEX

cars, 219–224
Cast Away (movie), 96, 133, 188
Castelo, Noah, 155
categorization, 151–154, 159–160
category conflict hypothesis, 152
causation, 183–188
    religion and, 194–195
Chalmers, David, 179–180
Chan, Alan, 215–216
Charged Lemonade, 210–211
Charkovsky, Igor, 44–47
Chaser the dog, 32–33
chatbots, 169–171. *See also* artificial intelligence (AI)
    Eliza effect with, 172–174
    as minds, 171–176
    origins of, 172–173
Chawarska, Katarzyna, 146–147
chimpanzees, 38, 59–60, 62
Chittka, Lars, 86
City Point, Virginia, 17–19
Civil War, US, 17–19, 240–241
Clasen, Mathias, 152, 162
Claude 3, 177–180, 243–244
Clinton, Bill, 228
Clinton, Paul, 150
clowns, 157–158
The Clueless, 230
CNN, 159
Coca-Cola, 205
cockroaches, 216
cognition
    in animals, 29–39, 41, 43–66, 74, 247–248
    comparative psychology on, 50–52
    dolphin-assisted births and, 44–47, 65–66
    ethology on animals, 57–59
    language and, 171–175
    mammalcentrism and, 83–87
    playfulness and, 76–79
    in raccoons, 107–108
cognitive appraisal process, 92–93
cognitive biases, facial recognition and, 81–83
cognitive dissonance, 240–241
cognitive ethology, 57–59
"A Cognitive Theory of Religion" (Guthrie), 193–194
Collider website, 144–145, 157
communication
    AI and, 171–176
    dog buttons for, 30–35
    narrating animals' thoughts and, 29–30
    referential, 32
Communications of the ACM, 165, 167
Companion Cube, 121–124, 126, 134
comparative psychology, 50–52, 53
    behaviorism and, 55–57
compassion, cuteness and, 29. *See also* empathy
competition, the problem of, 208–209

consciousness
    in AI, 176–182
    in animals, 29–39, 41, 43–66
    biology as prerequisite for, 179–180
    cognitive ethology on, 57–59
    Garland test for, 176–182
    mammalcentrism and, 84–86
    qualia and, 179
    testing for, 181–182
context, incorporation of in AI, 178–179
Coo, Joshua, 159
Copenhaver, Joby, 131
Cornerstone Church, 185
coulrophobia, 157–158
COVID-19, 187–188, 191–192
creepy counterfeits, 142–167
    Botox and, 158–160
    categorization and, 151–152
    clowns, 157–158
    dolls, 163–165
    fear response and, 151–157
    Humanity-Limiter and, 145, 146–147
    individual variability in experiencing, 154–155
    making them less creepy, 155–157
    masks, 160–163
    Muppets, 143–145
    puppets, 143–247
    robots, 165–167
    uncanny valley and, 147–163
Crick, Francis, 85
critical anthropomorphism, 59–60, 64–65
crocodilians, 67–74, 228
Cruz, Rubén, 230
CTI (*See* Canine Testicular Implantation)
culture, 14, 92
    food culture and, 103–105
    hunting and, 110–112
    Japan's cuteness culture, 112–113
    Jaws effect and, 106–108
    uncanny valley and, 155
Cunk, Philomena, 17
Current Affairs (TV show), 227–228
cuteness
    auditory, 24–25
    baby schema of, 20–25
    breeding animals for, 25–30
    cars and, 221–224
    environmental conservation and, 215–218
    Japanese culture and, 112–113
    playfulness and, 76–79
    psychopathology and, 102
    ugly animals and, 66–87

D

Dabiq Magazine, 228
Dacey, Mike, 38
Daggett, Cara, 224
Dahl, Roald, 67
Dale, Joshua Paul, 23, 76, 92, 112

287

INDEX

Dalí, Salvador, 205
Darwin, Charles, 8, 50–52, 81–82
dehumanization, 13–14, 231–240
   avoiding, 250
   biological roots of, 237–240
   conscious control over, 239–240
   everyday, 235–237
dehumanization hypothesis, 153
delayed-appraisal response, 92–93
detraction strategies, 231–235
Deutsche Welle (network), 217
Devine, Alexis, 30–31, 34
de Waal, Frans, 43, 59, 62
dice jail, 196–197
Diel, Alex, 167
Dinets, Vladimir, 71, 72, 73, 83–84
diseases, 190–193
Disney, 219–220
DOCTOR script, 172–173
Does the Dog Die? website, 134
dog buttons, internet, 30–35, 171–172
Dog Cognition Lab, 31
dolls, 163–165
Dolphin Communication Project, 43–44
dolphins
   human birth assisted by, 43–49, 64–65
   language in, 33
dopamine, 21
Dune 2 (movie), 234
Dungeons & Dragons, 196–197
Dunne, Joe, 3
Duolingo, 209

E

Earth First!, 218
effect, law of, 56–57
effectance motivation hypothesis, 10–11, 186
Eisnitz, Gail, 105
Elffers, J., 119
Eliza chatbot, 172–173
Eliza effect, 174, 175–176
emotional distance, 131–135
emotions
   of animals, 29–30
   car headlights and, 221–224
   emotional branding and, 210–211
   eyes and, 118
   false mirror problem with, 35–39
   fictional characters and, 131–135
   "Lamp" ads and, 202–204
   playfulness and, 78–79
   in sharks, 79–80
   Wally the alligator and, 67–74
empathy, 98–101
   artificial, 208–209
   cognitive appraisal and, 92–93
   cuteness and, 29
   definition of, 98–99
   dehumanization and, 231–235, 240

food culture and, 103–105
   harming animals and, 108–112
   hunting and, 109–112
   Lincoln and, 240–241
   psychopathy and, 101–103, 155
   slaughterhouse workers and, 105, 109, 126–127, 232
empiricism, 8
enemies, dehumanization of, 230–235
Energy (magazine), 147–148
environmental conservation movement, 211–212, 215–218
   petro-masculinity and, 223–224
Epley, Nicholas, 5, 12, 97–98, 109, 117, 231, 238
Euronews, 230
evolution
   anthropomorphism in science on, 50–52
   baby cuteness schema and, 22–25
   benefits of anthropomorphization and, 9–12
   dehumanization and, 237–240
   of doggy eyebrows, 35–37
   face recognition and, 93–95
   of fear responses, 90–91
   inanimate objects and, 138–139
   infant-directed speech and, 33–34
   intra-species recognition and, 6
   of language, 171–172
   religion and, 194–195
   uncanny valley and, 151–154
evolutionary aesthetics hypothesis, 151
Ex Machina (movie), 169, 176–177
experience, 7–8
   eyes, eyebrows, and, 28–29
   social motivation hypothesis and, 9–10, 186
The Expression of the Emotions in Man and Animals (Darwin), 50
Extraordinary Births (documentary), 48
eyebrows, 35–37
   dogs', 27–29
eyes
   breeding animals for cuteness and, 27–29
   on cars, 220–224
   cognitive appraisal and, 93–95
   crocodilians and, 75–76
   eye contact with animals and, 28
   inanimate objects and, 117–119
   as trigger, 13–14
Eyes Wide Shut (movie), 161
The Ezra Klein Show (podcast), 177

F

faces
   animals, babies, and, 20–25
   babies' appraisal of, 93–95
   Botox and, 158–160
   on cars and trucks, 220–224
   clowns and, 157–158
   dolls and, 163–165

INDEX

Kindchenschema for, 21–24
  masks and, 160–163
  pareidolia and, 118–119
  recognition of, 81–83
  on robots, 165–167
  uncanny valley effect and, 156–167
  wasps and, 81–83
Faces in the Clouds: A New Theory of Religion (Guthrie), 193–194
facial expressions, misinterpreting animals', 38–39
Fallout (TV show), 133–134
false mirror conundrum, 35–39
family background, 97
fast thinking, 93
fauxnads, 1–3, 15
fear grins, 38
fear responses
  to clowns, 157–158
  Jaws effect and, 106–108
  to spiders, 90–91
  uncanny valley and, 151–157
fictional characters, 131–135, 140
Finnish Hobbyhorse Championships, 128
Foggy Eye the shark, 79–80
Ford, John, 64
Ford F150 truck, 223–224
Forgetting Sarah Marshall (movie), 143
Fortuna, Tyfanee, 214–215
Fowler, Jeff, 156
Freymann, S., 119
funnel web spiders, 91

G

Gábor, Anna, 34
Gagarin, Yuri Alekseyevich, 226
Gagne, Patric, 103
Gaia, 218
Games Radar (magazine), 156
Garante per la protezione dei dati personali (GDPR), 170
Garland, Alex, 176–177
Garland test, 176–182
GDPR. *See* Garante per la protezione dei dati personali (GDPR)
gender
  naming objects and, 137
  naming storms and, 189–190
  vehicle "faces" and, 223–224
genocide, 232, 233–235
Gervais, Will, 186–187
Gibson guitars, 135–139
Gillette, Chris, 70–71
GLaDOS, 121–124
global warming, 218
god of the gaps, 186–187
Goldhagen, Daniel Jonah, 232
Gomez, Adrian, 187–188
Goodall, Jane, 59–60

Google, 181
Gordon, Rain, 140–141
Gore, Al, 218
Grant, Ulysses S., 17–19, 21, 240–241
Gray, Kurt, 7, 28–29, 152–153, 188, 195, 231
Greene, Joshua, 93
griefbots, 182–184
Griffin, Donald, 57–59, 84–85
Gross, Terry, 185
Grossman, Dave, 233
The Guardian (newspaper), 143
Gunkel, David, 198
Guthrie, Stewart, 116–117, 193–194, 195

H

HADD. *See* hyperactive agent-detection device (HADD)
Hagee, John, 185, 194–195
Hale, Edward, 54
Ham the chimpanzee, 38
Hanks, Tom, 96, 133, 188
happiness, marketing and, 205–206
Harrison, Benjamin, 228
Harrods, 164
Hattori, Jimmy, 213
Hauser, David, 192
headlights, 221–223
health monitor devices, 209
Heider, Fritz, 120
Heider-Simmel illusion, 120–121
Henney, Joie, 67–71, 73–74, 78–79
Herbie the Love Bug, 219–220
Herzog, Hal, 109
Hitler, Adolf, 220, 227
Hobbyhorse Revolution (documentary), 129
Hoover, Herbert, 228
Horowitz, Alexandra, 30, 31, 39
How Are You Peeling? Foods With Moods (Freymann & Elffers), 119
Hugo, Victor, 140
Humanity-Limiter, 129–130, 135
  AI and, 167, 172–173, 180
  categorization failures and, 151–154
  dehumanization and, 233–235
  dolls and, 163–165
  inanimate objects and, 139–142
  masks and, 162–163
  puppets and, 145, 146–147
  uncanny valley and, 149–150
humans
  creepy counterfeits of, 142–167
  definitions of, 6
  experience and agency in, 7–8
  face recognition and, 93–95
  minds of, 171–182
  playfulness as indicator of, 76–79
  threats to distinctiveness of, 208–209
  treating non-humans like, 6–7
  uncanny valley and, 147–160

INDEX

Hume, David, 8
The Hunchback of Notre Dame (Hugo), 140
hunters and hunting, 12, 109–112
Huron, David, 24–25
Hurricane Katrina, 184–186, 194–195
Huxley, Aldous, 225
hyperactive agent-detection device (HADD), 194–195
Hyson, Michael, 47

I

I Am Legend (movie), 134
IDAQ. *See* Individual Differences in Anthropomorphism Questionnaire (IDAQ)
Idriss, Shereene, 159
IEEE Spectrum (magazine), 148
Ig Nobel Prize, 2
IKEA, 201–204
imaginary friends, 130
inanimate objects, 115–142, 249
    Anthropo-Dial and, 125–135
    benefits of anthropomorphizing, 135–139
    caregiver response to, 121–124
    downside of anthropomorphizing, 139–142
    eyes and, 117–119
    fictional characters, 131–135
    Japanese culture and, 112–113
    Laundry Monkey, 115–117
    movement in, 119–121
    uncanny valley and, 147–163
An Inconvenient Truth (documentary), 218
incorporeal entities. *See also* artificial intelligence (AI)
    capitalism and, 197–199
    changing behavior based on, 188–193
    COVID ghosts, 187–188
    dice jail and, 196–197
    diseases, 190–193
    nature, 217–218
    religion, 193–196
    weather disasters, 184–186, 188–190, 194–195
Individual Differences in Anthropomorphism Questionnaire (IDAQ), 97–98
infant-directed speech, 33–34
in-group bias, 237–240
instinctive behaviors, 57–59
Insurance Institute for Highway Safety, 223
intelligence. *See also* cognition
    in crocodilians, 74
    in raccoons, 107–108
intentionality, 10–11
    clowns and, 157–158
    in weather disasters, 184–186, 194–195
International Criminal Tribunal, 234
International Journal of Hospitality Management, 229
ISIS, 228
isolation, 122
It (King), 158
Italian Data Protection Authority, 170

J

Jackson, Andrew, 228
Japan
    kawaii culture in, 112–113, 155
    mascot culture in, 212–213
    raccoons in, 216–217
Jaws (movie), 106–108
jingoism, 12
John Wick (movie), 134
Jonze, Spike, 202, 204

K

Kahneman, Daniel, 93
kawaii culture, 112–113
Kennedy, Caroline, 226
Kennedy, Jacqueline, 226
Kennedy, John F., 226–227
Kermit the Frog, 144
Kershenbaum, Arik, 32
Khrushchev, Nikita, 226–227
Kindchenschema, 21–24, 27
    cars and, 222
    cuteification of ugly animals and, 75–76
    eyebrows and, 28
    mascots and, 213
    politics and, 229–231
    psychopathology and, 102
    puppets and, 145
    robots and, 166
King, Martin Luther Jr., 9–10
King, Stephen, 158
Kitzinger, Sheila, 47
Kjeldgaard-Christiansen, Jens, 152, 162
Koch, Christof, 85
Kodak, 210
Koko the gorilla, 33
Kreutinger, Adam, 145
Kringelbach, Morten, 21, 29
Kugler, Logan, 165

L

Lacroix, Guy, 152
Lalonde, Jess, 99–101
LaMDA. *See* Language Model for Dialogue Applications (LaMDA)
"Lamp" commercial, 202–204
language
    AI and, 171–182
    animals and, 30–35
    consciousness and, 84–85
    dehumanization with, 234–235
    dolphins and, 46–47
    marketing and, 207–209
    purpose of, 171–172
    religion and, 195

INDEX

stochastic parrots and, 175–176
  as trigger, 13–14
Language Model for Dialogue Applications (LaMDA), 181
LAOM. *See* levator anguli oculi medialis (LAOM)
large language models (LLMs), 171, 175. *See also* artificial intelligence (AI)
  griefbots, 182–184
  literature reviews and, 243–244
Lasseter, John, 204
Laundry Monkey, 115–117, 244
law of effect, 56–57
learning, in animals, 56–57
Lee, Robert E., 17, 19
Lemoine, Blake, 181
levator anguli oculi medialis (LAOM), 27–28
Lewes, George Henry, 8
Liam chatbot, 169–171
Lilly, John, 46–47
Lincoln, Abraham, 17–19, 21, 25, 228, 240–241
Lombrozo, Tania, 187
loneliness, 96–97, 188
  dehumanization and, 238–239
  inanimate objects and, 122
Lorenz, Konrad, 21–22, 23–24, 57
Lorimer, Jamie, 212
The Love Bug (movie), 219–220
Loy, Loredana, 134–135
Luca Inc., 170
Luxo Jr. (movie), 204

M

M3GAN android, 147, 150, 153, 156, 157, 167
MacDorman, Karl, 165, 236
Machiavellianism, 102
Mae West bottle, 205
Major League (movie), 96
Making Monsters: The Uncanny Power of Dehumanization (Smith), 232
mammalcentrism, 83–87
Man and Dolphin (Lilly), 46–47
marketing, 14–15, 201–224
  car "faces" and, 220–224
  Disney and, 219–220
  emotional branding and, 210–211
  environmental conservation and, 215–218
  IKEA, 201–204
  mascots and, 211–215
  negative consequences of anthropomophization in, 207–209
  self-awareness and, 244–245
  sex and, 205–207
Marsh, Amy, 141
The Martian (Weir), 115
mascots, 14, 211–215
masks, 160–163
Mathieu, Christian, 202

Matsui, Ichiro, 213
McCartney, Margaret, 192
McDonald's, 210
McGill, Ann, 190
McLachlan, Sarah, 133
McLeod, Lianne, 217
meaning, making, 118–119
Metro (newspaper), 141
Mi'kmaq people, 110–112
Miller, Gregg, 1–3, 15
The Mind Club (Wegner & Gray), 7
Mollick, Ethan, 177
morality, 7–8, 198–199
  brachycephalic animal breeds and, 26–27
  cognitive appraisal and, 93
  dehumanization and, 231–240
  fiction characters and, 131–135
  food culture and, 103–105
  harming animals and, 108–112
  inanimate objects and, 136–137
  marketing and, 207
  masks and, 160–163
  propaganda and, 224–241
Moral Machines (Wallach & Allen), 183
Morgan, Conwy Lloyd, 52–55
Morgan's canon, 50–53, 57, 58
Mori, Masahiro, 147–148, 165, 236
mortality salience hypothesis, 151
movement, as trigger, 13–14
  animal breeding for cuteness and, 27–29
  inanimate objects and, 119–121
Muppets, 143–145
mutuality, 9–10

N

names and naming, 59–60
  of cars, 220–221
  of hurricanes, 188–190
  inanimate objects and, 136–137
  marketing and, 205
Nao robot, 165–166
narcissism, 102
National Lampoon's Christmas Vacation (movie), 134–135
National Oceanic and Atmospheric Administration (NOAA), 189
nature, 217–218
Nature (magazine), 51, 52
Nazism, 23–24, 220, 227
Nelli, Kim, 48
neurobiology
  car headlights and, 221–222
  consciousness and, 85–86
  eyes and, 117–119
  face recognition and, 93–95
  movement and, 119–121
  psychopathology and, 101–103, 155
Neuticles, 1–3, 15
Newsweek (magazine), 5

INDEX

New York Declaration on Animal
  Consciousness, 85
New York Times (newspaper), 47, 202, 203
Nietzsche, Friedrich, 186–187
Nikoll, Gabby, 70–71
NOAA. *See* National Oceanic and
  Atmospheric Administration (NOAA)
Nombre, Rosalie, 230
NPR, 185

O

Obama, Barack, 228
objectophilia, 139–141, 247
O'Daniel, Gerald, 159
Ologies (podcast), 234
On Killing: The Psychological Cost of Learning to Kill in War and Society (Grossman), 233
Optimus robot, 165
Orca Behavior Institute, 63
orcas, attacking, 62–64
O'Reilly, Terry, 201
Origin of Fear of Clowns Questionnaire, 158
overactive anthropomorphizers, 96–98
oxytocin, 28
Ozhiganova, Anna, 46

P

pain
  babies' experience of, 37–38
  chronic, responses to, 192
  racism in treatment of, 236–237
panda bears, 211–212
Panera Bread, 210–211
parasocial relationships, 244–247
pareidolia, 118–119, 166, 221–223
Park, Gene, 156–157
PARO, 165–167
parsimony principle, 61–62
Pascal's Wager, 194
pathogen avoidance hypothesis, 151
Pavlov, Ivan, 56
Pepin-Neff, Chris, 106
personality
  in AI, 169–171
  Anthropo-Dial and, 125–135
  empathy and, 98–101
  idiosyncrasies in anthropomorphism and, 92–96
  masks and, 161–163
  overactive anthropomorphizers and, 96–98
  psychopathology and, 101–103, 155
  in ugly animals, 71
  uncanny valley and, 154–155
petro-masculinity, 224
pets. *See* animals
Phillips, Maya, 161
Pier 1, 201, 202
Pilley, John W., 32–33
Pixar Animation Studios, 204

plague doctor masks, 160–161
Plancarte, Kristyn, 34
plastic surgery, 159–160
play, 14
  in animals, 72–74, 76–79
  definition of, 72–73
  hobbyhorsing, 127–129
Plotnik, Joshua, 64
Pocho the crocodile, 71–72
Poetry of America (Dalí), 205
The Polar Express (movie), 150, 154, 156
politics, 14–15, 198–199, 224–241
  dehumanization in, 231–240
  Kindchenschema faces and, 229–231
  petro-masculinity and, 223–224
  pets in, 225–229
Poltergeist (movie), 157
Pooley, Terry, 136–137
Portal (video game), 121–124, 126, 134
Porter, Horace, 18–19
The Power of Human (Waytz), 7
Prithvi, 218
problem of competition, 208–209
problem-solving, 107–108
propaganda, 14–15, 224–241
  dehumanization in, 231–240
  everyday dehumanization and, 235–237
  use of animals in, 225–229
  warfare and, 232–235
psychological traits, 96–98
psychopaths, 101–103, 155
Puppet Nerd, 145
puppets, 143–147
purrs, 25
Putin, Vladimir, 228

Q

qualia, proving the existence of, 179
The Question of Animal Awareness (Griffin), 58

R

raccoons, 107–108, 216–217
Racine Zoo, 216
racism, 12, 233–235, 236–237
Rando, Carolyn, 40
RAOL. *See* retractor anguli oculi lateralis (RAOL)
Rascal Racoon (TV show), 216–217
Reagan, Ronald, 228
Reardon, Sara, 40
Reborn dolls, 164–165
referential communication, 32
Reinboth, Tim, 182–183
religion and spirituality, 14–15, 193–196
  god of the gaps and, 186–188
  Japanese cuteness culture and, 112–113
  natural disasters and, 185–186
  uncanny valley and, 154
Replika app, 169–171

INDEX

retractor anguli oculi lateralis (RAOL), 28
Reuters, 164, 165
Robinson, Margaret, 110–111
Robinson, Nathan J., 227–228
Robotics and Automation Society International Conference on Humanoid Robots, 148
robots, 14–15
    attitudes toward, 166–167
    bomb-disposal, 138–139
    M3GAN android, 147, 150, 153, 156, 157, 167
    reducing the uncanny valley in, 155–156
    uncanny valley and, 147–150, 165–167
Robson, David, 190
role-play, 130
Romanes, George, 50–52
romantic and sexual relationships
    with AI, 169–171
    alligators and, 72
    in-group bias and, 238
    with objects, 139–141
    sadism and, 103
Roomba, 154
Rossellini, Isabella, 89
Rossiter, John, 210
Royal College of Veterinary Surgeons, 2
Rwandan genocide, 232, 234

S

sadism, 103
Safina, Carl, 60
Samuelson, Alexander, 205
SAPAN. *See* Sentient AI Protection and Advocacy Network (SAPAN)
Sartre, Jean-Paul, 31
Sarvary, Miklos, 155
Schein, Martin, 54
Schoenherr, Jordan, 152
Schroll, Roland, 207
science, 4–5
    anecdotal anthropomorphism in, 50–53
    on animal communication, 31–35
    on babies' experience of pain, 37–38
    behaviorism in, 55–57
    comparative psychology, 50–52, 53
    critical anthropomorphism in, 59–60, 64–65
    criticisms of anthropomorphism in, 8–9
    divisions of anthropomorphism in, 60–62
    ethology in, 57–59
    mammalcentrism in, 83–86
    Morgan's canon in, 50–53
    natural selection and, 50–52
    on orcas attacking boats, 62–64
    parsimony principle in, 61–62
    rich *vs.* lean interpretations in, 61–64
    utility of anthropomorphism in, 44–62
Science (magazine), 40
Science Friday (podcast), 198
scorpion suicides, 51–52

Segel, Jason, 143–145
Sentient AI Protection and Advocacy Network (SAPAN), 183–184
Serpell, James, 12, 26, 27
Seth, Anil, 176–177, 180
Seward, William, 18
sharks, 65–66, 79–80, 106–108
Shavitt, Sharon, 190
Shedden, Gilberto "Chito," 71–72
Shepard, Dax, 23
Sherman, William T., 18
Shi, Victoria, 230
Shields, Monika Wieland, 63, 64
Short, Kevin, 217
Sidenbladh, Erik, 45
Simmel, Marianne, 120
Siri, 198
Sirius Institute, 47, 48
Skinner, B. F., 56
Skuse, Alanna, 193
Slaughterhouse (Eisnitz), 105
slaughterhouse workers, 105, 109, 126–127, 232
sloths, 23, 121
slow thinking, 93
Smith, David Livingstone, 151–152, 157, 232
Smokey Bear, 212
social connections
    AI and, 169–199
    empathy and, 99–103
    eyes, eyebrows, and, 28–29
    harnessing anthropomorphism for, 243–251
    human drive for, 9–10
    marketing and, 205–206
    natural disasters and, 186
    parasocial relationships, 244–247
    with ugly animals, 70–74
social motivation hypothesis, 9–10
social robots, 165–167
Society for the Prevention of Cruelty to Animals (SPCA), 214
Sociopath: A Memoir (Gagne), 103
sociopathy, 102–103
SoftBank, 165–166
solicitation purrs, 25
Some We Love, Some We Hate, Some We Eat: Why It's So Hard to Think Straight About Animals (Herzog), 109
Sonic the Hedgehog, 156–157
sounds, as triggers, 24–25
space race, 225–227
SPCA. *See* Society for the Prevention of Cruelty to Animals (SPCA)
spiders, 81, 89–91, 92, 93, 96–97, 117, 248–249
Spielberg, Steven, 107
spirituality. *See* religion and spirituality
Spot robot, 166
Stanton, Lauren, 107–108
stochastic parrots, 175–176

INDEX

Stoller, Nick, 144
Suffering Index, 195
supernatural, 186–188. *See also* religion and spirituality
Swift, Kim, 122, 123

T
The Telegraph (newspaper), 48
Tenney, John E. L., 187
Tesla, 165
testicles, prosthetic dog, 1–3
theory of mind, 10–11
　AI and, 171–182
　animals and, 29–39, 247–248
　face recognition and, 93–96
　uncanny valley and, 152–153
Thinking Dog Center, 31
Thomson, Allen, 51–52
Thorndike, Edward, 55–56
thought narration, 29–30
threat ambiguity theory, 151–152
Thus Spoke Zarathustra (Nietzsche), 186–187
timescale bias, 121
Tinbergen, Niko, 57
Tonetti, Elena, 47
Torchinsky, Jason, 223–224
Touré-Tillery, Maferima, 190
Toy Story 2 (movie), 132–133, 204
tribalism, 12
triggers, 12–15, 116–117
　caregiver response to babies, 20–25
　sounds, 24–25
　uncanny valley, 157–158
Turing, Alan, 174
Turing test, 174–175
turkeys, mating behavior of, 53–55
turtles, 76

U
UFOs, 120
ugliness, 66–87, 248–249
　cuteification of, 75–76
　mammalcentrism and, 83–87
　playfulness and, 76–79
　sharks and, 79–80
　wasps and, 81–83
Unböring ad campaign, 202–204
uncanny valley effect, 147–163
　ableism and, 236
　Botox and, 158–160
　clowns and, 157–158
　dolls and, 163–165
　faces and, 156–167
　masks and, 160–163
　reducing, 155–157
　robots and, 165–167
　variability in experiencing, 154–155
Undark (magazine), 182–183
United Nations Genocide Convention, 234

unpredictability
　creepiness and, 151, 158
　effectance motivation hypothesis on, 10–11
　of movement, 119–121
　weather disasters and, 183–186
US Forest Service, 212–213
Uysal, Ertugrul, 208

V
Veterinary Council of New Zealand, 2–3
Volkswagen Beetle, 220
von Frisch, Karl, 57

W
waggle dance of bees, 57
Wallach, Wendell, 183
Wallius, Maisa, 129
Wally the alligator, 67–71, 72–74, 86–87
　cuteification of, 75–76
　playfulness of, 76–79
Walsh, Bill, 220
Wan, James, 157
Wang, Lili, 190
Ward, Alie, 234
warfare, 233–235
Washington Post (newspaper), 21, 31, 32, 34, 156, 181
Water Babies (Sidenbladh), 45
water births, 43–49
Watts, Peter, 171–175
Waytz, Adam, 5, 7, 97–98, 109, 238
weather disasters, 184–186, 188–190, 194–195
Wegner, Daniel, 7, 28–29, 152–153, 195, 231
Weigelt, Sarah, 167
Weighted Companion Cube, 121–124, 126, 134
Weir, A., 115
Weizenbaum, Joseph, 172–173, 175–176
Whitmire, Steve, 144
Why Animals Talk (Kershenbaum), 32
Wolpaw, Erik, 122
Woodland Park Zoo, 216
World Meteorological Organization, 189
World Wildlife Fund (WWF), 211–212
Wright, Kailin, 131–132
Wu, Katherine J., 64
Wynne, Clive, 31–32, 60–62

X
Xenophanes, 8

Y
Yang, Lin, 205, 209, 210–211, 244–245
yellow fever, 190–191

Z
Zenato, Christina, 79–80

ABOUT THE AUTHOR

Dr. Justin Gregg is a Senior Research Associate with the Dolphin Communication Project and adjunct professor at St. Francis Xavier University, where he lectures on animal behavior and cognition. He received his PhD from the School of Psychology in Trinity College, Dublin, having studied dolphin social cognition. He is the author of the bestselling book *If Nietzsche Were a Narwhal: What Animal Intelligence Reveals About Human Stupidity*. Justin currently lives in rural Nova Scotia, where he writes about science and contemplates the inner lives of the crows that live near his home.

RAISING READERS
Books Build Bright Futures

Thank you for reading this book and for being a reader of books in general. As an author, I am so grateful to share being part of a community of readers with you, and I hope you will join me in passing our love of books on to the next generation of readers.

Did you know that reading for enjoyment is the single biggest predictor of a child's future happiness and success?

More than family circumstances, parents' educational background, or income, reading impacts a child's future academic performance, emotional well-being, communication skills, economic security, ambition, and happiness.

Studies show that kids reading for enjoyment in the US is in rapid decline:

- In 2012, 53% of 9-year-olds read almost every day. Just 10 years later, in 2022, the number had fallen to 39%.
- In 2012, 27% of 13-year-olds read for fun daily. By 2023, that number was just 14%.

Together, we can commit to **Raising Readers** and change this trend. How?

- Read to children in your life daily.
- Model reading as a fun activity.
- Reduce screen time.
- Start a family, school, or community book club.
- Visit bookstores and libraries regularly.
- Listen to audiobooks.
- Read the book before you see the movie.
- Encourage your child to read aloud to a pet or stuffed animal.
- Give books as gifts.
- Donate books to families and communities in need.

Books build bright futures, and **Raising Readers** is our shared responsibility.

For more information, visit **JoinRaisingReaders.com**

Sources: National Endowment for the Arts, National Assessment of Educational Progress, WorldBookDay.org, Nielsen BookData's 2023 "Understanding the Children's Book Consumer"